Studies in Choice and Welfare

More information about this series at http://www.springer.com/series/6869

William V. Gehrlein • Dominique Lepelley

Elections, Voting Rules and Paradoxical Outcomes

Springer

William V. Gehrlein
Department of Business
Administration
University of Delaware
Newark, Delaware
USA

Dominique Lepelley
University of La Réunion
Saint-Denis, Ile de La Réunion
France

ISSN 1614-0311 ISSN 2197-8530 (electronic)
Studies in Choice and Welfare
ISBN 978-3-319-64658-9 ISBN 978-3-319-64659-6 (eBook)
DOI 10.1007/978-3-319-64659-6

Library of Congress Control Number: 2017951503

Printed on acid-free paper

This Springer imprint is published by Springer Nature
The registered company is Springer International Publishing AG
The registered company address is: Gewerbestrasse 11, 6330 Cham, Switzerland

To Peter C. Fishburn: An outstanding scholar, mentor, and friend
William V. Gehrlein

À Maurice Salles pour son exemple, et à Françoise pour son soutien
Dominique Lepelley

Preface

The topics of Election Procedures and Voting Paradoxes have been formally studied by a very large number of philosophers, mathematicians, economists, and political scientists since the groundbreaking work in those areas by the remarkable eighteenth-century French scholars Borda and Condorcet. It was Condorcet who noted that it was not adequate to simply identify possible election outcomes and voting paradoxes that could happen. He stressed that a study of the probabilities that such events might occur is critical to evaluating their potential impact in actual elections. A huge effort has been expended since the days of Condorcet's early work to obtain probability estimates for the likelihood that such outcomes will be observed in election settings. The vast majority of the theoretical work in this area has taken place in the last 50 years, and the level of mathematical sophistication that drove those efforts escalated significantly as that work progressed.

The book *Voting Paradoxes and Group Coherence: The Condorcet Efficiency of Voting Rules* was published in 2011 as a result of our desire to provide a technical summary of the advancements that had been made in these theoretical studies up to that time. Soon after completing that book, we concluded that it would be a very worthwhile endeavor to rewrite much of the contents of that book in a less technical format, in order to disseminate those results to a much wider audience, and to discuss the practical implications of that work. It also seemed to us at the time that this would be a rather easy task to complete.

The reality of producing such a descriptive study turned out to be far more complicated than we anticipated. Many new researchers suddenly entered into that area of research, and they were producing some extraordinary new results by using very sophisticated techniques. This led us into a series of new studies with a number of different coauthors that greatly extended what was known about the theoretical probabilities of observing election outcomes and voting paradoxes. A very important opportunity also presented itself for us to do a series of studies with Florenz Plassmann that provided solid empirically based evidence to support the general conclusions that the theoretical studies had generated regarding the Condorcet efficiency of voting rules.

After 6 years of additional work, we now present this study that describes the theoretical work that has been done regarding the likelihood that election outcomes and voting paradoxes will be observed in a much less technical format. An emphasis is placed both on the extensive practical implications of this work and on a thorough discussion about the empirically based research that has been performed to provide sound support for many of the conclusions that have been developed with that theoretical analysis.

On a personal note, Gehrlein wishes to extend sincere gratitude to his wife Barbara Eller, who has been extremely supportive and encouraging through the long course of this project. The list of coauthors who were involved in all of the studies that led to this book is both long and impressive, and sincere gratitude is extended to each of them for their willingness to use their extraordinary talents to solve many very difficult problems. Dominique Lepelley stands out from this list to deserve particular recognition, since he was completely responsible for orchestrating the efforts that significantly extended the scope of this work well beyond the point when it seemed that all limits of computational possibilities had already been reached. Finally, gratitude is extended to Peter Fishburn for introducing all of this in a classroom at Pennsylvania State University in 1971 and for presenting the challenge to develop some probability representations for the likelihood of observing Condorcet's Paradox. That assignment has now been completed with this final project, after 46 years of work!

Lepelley expresses his gratitude to Maurice Salles, who suggested in 1974 that he should write his research paper on the probability of observing Condorcet's Paradox. That paper marked the beginning of an exciting scientific adventure, which this book concludes. Many thanks are extended to the numerous coauthors who were involved in this story: Nicolas Andjiga, Fabrice Barthélémy, Virginie Béhue, Sven Berg, Gwenaël Bily, Franck Bisson, Jean Bonnet, Frédéric Chantreuil, Olivier de Mouzon, Abdelhalim El Ouafdi, Pierre Favardin, Marc Feix, Erik Friese, Thibault Laurent, Michel Le Breton, Ahmed Louichi, Antonin Macé, Mathieu Martin, Boniface Mbih, Vincent Merlin, Issofa Moyouwou, Michel Paul, Patrick Pierron, Florenz Plassmann, Jean-Louis Rouet, Nicolas Sauger, Achill Schürmann, Jérôme Serais, Hatem Smaoui, Fabrice Valognes, Laurent Vidu, ..., and of course Bill Gehrlein, with whom 28 papers have been cowritten!

Newark, Delaware, USA
La Réunion Island, France
May 2017

William V. Gehrlein
Dominique Lepelley

Contents

List of Abbreviations

ANES	American National Election Studies dataset
AV	Approval voting
BER	Borda elimination rule
BR	Borda rule
BRR	Borda rule runoff voting rule
CL	Condorcet loser
CR	Copeland voting rule
CW	Condorcet winner
EIAC	Extended impartial anonymous culture condition
EIC	Extended impartial culture condition
DC	Dual culture condition
EBR	Extended Borda rule
ENPR	Extended negative plurality rule
EPR	Extended plurality rule
ERS	Electoral reform society dataset
EV	Three-scale evaluative voting
EWSR(λ)	Extended weighted scoring rule with weights $(1, \lambda, 0)$
FIAC	Forced ranking impartial anonymous culture condition
FIC	Forced ranking impartial culture condition
IAC	Impartial anonymous culture condition
IC	Impartial culture condition
IRV	Instant runoff voting rule
MIAC	Modified impartial anonymous culture condition
MIC	Modified impartial culture condition
NPER	Negative plurality elimination rule
NPR	Negative plurality rule
NPRR	Negative plurality rule runoff voting rule
P-E	Polya–Eggenberger Urn model
PB	Politbarometer dataset
PER	Plurality elimination rule
PMR	Pairwise majority rule

PR	Plurality rule
PRR	Plurality rule runoff voting rule
RV	Range voting rule
UC	Uniform culture condition
WSR(λ)	Weighted scoring rule with weights $(1, \lambda, 0)$

Chapter 1
Elections and Voting Paradoxes

1.1 Introduction

People have long been interested in the process of how groups of individuals should go about selecting a winner from a set of available alternatives. At the simplest level, this could consist of a group of children who are trying to determine which particular game they should play together on a summer afternoon. At the opposite extreme, it could be the entire population of a country that is attempting to determine who should be chosen as their leader during a period of crisis. Both of these scenarios and everything on the spectrum between them can be viewed in the context of an election. The individuals who are making the decision are the voters of the electorate, the available options are the candidates and the selected option is the winner of the election.

If each of the voters happens to have the same most preferred candidate in a given election, then the determination of the winner is a trivial task. That most preferred candidate is selected as the winner, and every voter will be happy to obtain their most preferred outcome. It is unfortunately immensely more likely to have a scenario in which there is some disagreement among the voters regarding which of the candidates is thought to be best. Once this situation arises, each of the individual voters cannot get what they most prefer as an outcome, so the determination of how to decide which candidate should be chosen as the winner becomes an important issue.

When there are only two candidates to choose from, it is almost certain that the idea of majority rule will be suggested as the voting rule, so that the candidate that is more preferred by the greater number of voters will be the winner. A sense of fairness suggests that this candidate should be elected. Rousseau (1762) presents a classic analysis of the issue of the basic fairness of majority rule voting by suggesting that the *general will* of the electorate should always serve as the legitimate norm for making decisions in any group. Any particular member of the electorate who is voting in an election is then viewed in the context of trying to

W.V. Gehrlein, D. Lepelley, *Elections, Voting Rules and Paradoxical Outcomes*,
Studies in Choice and Welfare, DOI 10.1007/978-3-319-64659-6_1

decide which of the two candidates conforms best to that *general will* of the electorate. Those individuals who vote for the candidate that is not elected are then simply viewed as being incorrect in their own assessment of which candidate is most in conformance to the *general will*.

As with a large majority of philosophical arguments, this fundamental belief in the overall fairness of majority rule is not a universally held principle. Arguments that are posed against using majority rule typically attack it because it ignores some possible issues that can arise regarding the intensity of preferences of the voters. This was formally argued by Don Joseph Isadore Morales in a paper that is summarized in Daunou (1803). The primary argument that Morales presented was based on the notion that situations could exist in which there is a minority group of voters who have a very strong preference that one of the two candidates should be adopted, while the majority of voters prefer the other candidate, but are only marginally opposed to the one that is preferred by the minority. If the sizes of the two voting groups are nearly equal, Morales suggests that the strong preference of the minority in this case should outweigh the weak preference of the majority opinion. Voting procedures would therefore have to ask the individual voters to report some measure of their relative degrees of preference for the candidates, as opposed to simply asking which of the two candidates is more preferred.

Daunou (1803, p. 244) opposes Morales' arguments by drawing an analogy between voters having intense preferences and having a strong will. He notes that society may have benefited on occasion from acts of strong-willed people, but that there are "a great number of infamous disasters" that can be attributed to them, so that all votes must be treated with equal value, regardless of the degree of intensity of preference that might be associated with them. The fact that issues that are related to elections have been analyzed for a very long time is illustrated by a reference (Gaertner 2005, p. 235) to comments made by Pliny the Younger of Ancient Rome regarding the inevitability that all votes must be counted equally for all voters, despite the added apparent criticism that "nothing is so unequal as the equality" of this practice.

As with most philosophical arguments, much has been written about this topic since the early work that has been cited, but the argument regarding the necessity to consider the strength of voters' preferences has not been resolved (see Gehrlein 2006). We generally follow the same direction as most of the work in this area and ignore intensity of voters' preferences as an issue, but some aspects of the notion of intensity of preference will be addressed in the final chapter. So, all voters have the same input into the voting process, and no subgroup of voters within the electorate will have any more influence on the outcome of the voting process than any other subgroup of the same size. This does not rule out the possibility that some individuals might be more persuasive than others in arguing to support their particular candidate of choice during any preliminary debate to compare the candidates. However, once each of the individual voters has determined their particular preferences on the candidates, each of them will then have exactly the same influence on the election outcome.

We have seen that commentary regarding how elections should be conducted can been traced back to the time of Ancient Rome, and it is not surprising to learn that the topic has been considered by many scholars since then. McLean (1990) discusses observations of Ramon Lull from the fourteenth century and of Nicolas Cusanus from the fifteenth century that are related to voting procedures. Similarly, Lagerspetz (1986) and Gaertner (2005) present observations of Pufendorf from the seventeenth century. All of the concepts that were considered in these studies are relevant to issues that are related to conducting elections, but the first known formal mathematical analysis of this problem was conducted nearly 250 years ago by two French mathematicians named Jean-Charles de Borda (Chevalier de Borda) and Marie Jean Antoine Nicolas de Caritat (Marquis de Condorcet).

Their work was based on the fact that things become much more complicated for elections with more than two candidates. When it is asked how a winner should be elected in this case, it can be expected that an appeal for majority rule will be heard again, and Borda and Condorcet both found that a difficulty now arises because there are two ways to define what majority rule might mean when there are more than two candidates. Borda and Condorcet were very different individuals, and their work on the mathematical analyses of elections and voting rules formed the foundation for a line of research that still continues. Before we delve into considering exactly what these two mathematicians wrote about elections and voting rules, we take a brief look at the historic events that were happening in the world around them as their work continued, and at the interaction that took place between these two contemporaries.

1.2 A Tale of Two Mathematicians

As you might deduce from the fact that the title of this section is a parody of the title of Charles Dickens' novel *A Tale of Two Cities*, the work of Borda and Condorcet that was related to the analysis of elections and voting rules was completed in the period leading up to and during the French Revolution, and we shall see why all of this work ended abruptly in 1794. It will also be seen that both Borda and Condorcet would very likely still have been famous, even if they had never written a single word about elections.

Borda was born in the city of Dax in 1733 and he became an officer in the French Navy, after first entering the army at a young age. His duty as a naval officer was primarily involved with serving as a military engineer. As a result of his work and the treatise that evolved from it on the movement of projectiles from artillery, he was elected as a member of the French Academy of Sciences in 1764. Borda's continuing military service called for him to make a number of crossings of the Atlantic Ocean between 1765 and 1775 for scientific studies and military purposes. The American Revolution began in 1775 and France allied with America in 1778 in their fight against the British. Borda became heavily involved in this action as a result of being a naval officer, since he was placed in command of several ships that were part of the French Fleet in the Caribbean Ocean, near the American coastline. After being successfully involved in several important battles, Borda was captured and taken as a prisoner of war, but he was soon released on a pardon and returned to France. He then continued to invent instruments that were related to navigation, was involved in the modernization of the French Fleet and worked on numerous aspects in the establishment of the metric system.

Borda was very well established as a major contributor to the field of engineering in France. Almost every person who has visited Paris has seen the Eiffel Tower, and it is remotely possible that some might not have noticed the fact that the names of 72 people are engraved on the faces of the tower just below the first balcony. These names represent people who were among the most prominent mathematicians and engineers in France up to the time that the tower was built. Among the names listed is that of Borda. You might very well be asking at this point exactly what Borda did that had anything to do with elections and voting rules, and the answer is that buried among the huge quantity of output from his work lies one single paper that he wrote about conducting elections to develop a voting rule that he called *Election by Order of Merit*. His presentation of this study was made to the French Academy of Sciences on June 16, 1770, but it was not published in the proceedings of the Academy until 1784; and the long delay between presentation and publication will be addressed later. An extensive biography of Borda was published by Mascart in 1919 [Reprinted in Mascart (2000)] and fewer than five pages of this 800 page tome are dedicated to Borda's 1784 paper on elections, to indicate very strongly that this single paper played a small role in the totality of Borda's accomplishments, while it still had a highly significant impact on the study of election procedures and placed everything that follows into motion.

After seeing Borda's name on the Eiffel Tower, you can follow the sidewalk of the street that runs beside the Seine, with the river on your left hand side as you proceed, and after a good walk approach a large bronze statue. That statue was erected in honor of the Marquis de Condorcet, who was a very famous philosopher and mathematician. Condorcet was born in 1743 in Ribemont and he was educated at the Jesuit College in Reims. He later demonstrated very promising talent as a mathematician at the College of Navarre in Paris, and was elected as a member of the French Academy of Sciences in 1769, where he became a permanent secretary in 1777. Condorcet wrote many papers and books on a number of topics, but generally in the field of mathematics. One of Condorcet's major accomplishments came when he published his *Essay on the Application of Analysis to the Probability of Majority Decisions* in 1785. This book made a highly significant contribution by introducing the concept of analyzing election outcomes on the basis of the probability that they might occur, and it was very solidly based in mathematical analysis.

Condorcet received great regard from his contemporaries for introducing the mathematical precision of probability into the process of decision making. His fame in this endeavor is indicated by the fact that he is mentioned several times in another famous novel that is set during the French Revolution, Victor Hugo's well-known *Les Misérables* from 1862. One quote that specifically makes reference to

Condorcet's precision in measuring his probabilities and human behavior in general is [Translation from Wilbour (1909, p. 568)]:

> All sublime conquests are, more or less, the prizes of daring. In order that the Revolution should take place, it does not suffice that Montesquieu should forsee it, that Diderot should preach it, that Beaumarchais should announce it, that Condorcet should calculate it, that Arouet should prepare it, that Rousseau should premeditate it; it is necessary that Danton should dare it.

Condorcet's 1785 treatise formed the basis of competing theories between Condorcet and Borda, and it was quite critical of the conclusions that were reached and of the voting rule that was proposed in Borda's presentation to the Academy in 1770. So, why was Borda's presentation from 1770 not published in the proceedings of the French Academy of Sciences until 1784? Most writers are very familiar with publication delays, but this example is definitely far beyond extreme. McLean (1995) has suggested that it is very likely that Borda played little or no role in the process of having the paper from his presentation published. It was in fact Condorcet who used his position in the Academy to add Borda's 1770 paper to the Proceedings of the French Academy of Sciences in 1781, which was not actually published until 1784. Any reader of Borda's paper would therefore be left with the impression that his work was completed much later than when it actually was presented before the Academy. This would make Borda's work appear more current and relevant for criticism in Condorcet's 1785 book. Condorcet added that he had heard of Borda's earlier results, but that he was unaware that anything had been written on the topic. However, Condorcet provided written comments on Borda's work in 1784, stating (McLean 1995, p. 16) that the results were both "very important" and "totally original."

It is not possible to know with certainty what the personal relationship between Borda and Condorcet was like, but there is some good evidence that things had deteriorated badly between the two. The strong criticism of Borda's work in Condorcet's 1785 book took a more personal and antagonistic tone later on. Condorcet wrote of Borda's focus on engineering and made reference to him as "having abandoned mathematics for petty applied science" (Baker 1975, p. 42). Particularly harsh words from an academic perspective come when Condorcet wrote that Borda had written papers "that nobody has ever spoken of ... or ever will" (McLean and Hewitt 1994, p. 9). Young (1995) describes Condorcet's comments about Borda as indicating "a certain amount of personal venom." Baker (1975, p. 42) also reports that a dispute arose between the two as a result of an attempt by the famous mathematician d'Alembert to obtain some financial payment to support Condorcet at the time when Condorcet was appointed as Assistant Secretary of the French Academy of Sciences.

A totally different version of the nature of the relationship between Borda and Condorcet is described by Black (1958, p. 179), who writes that they remained close friends throughout their lives. This notion is partly supported by comments in Mascart (2000) to indicate that Borda very much appreciated Condorcet and remained close to him. This is indicated by the observation that Borda and

Condorcet worked on the same problem of hydrodynamics between 1770 and 1780, and that work resulted in a disagreement between them. In a letter to Condorcet, Borda wrote: "I am upset that we disagree, but it will not prevent both of us from going into Paradise ... I kiss you from my heart." Both Borda and Condorcet also served on the Committee on the Units of Measurement in 1790 with Lagrange, Laplace and Monge to show that both were considered to be comparable to these very notable mathematicians of that period.

Condorcet was a very progressive thinker who went on to write in support of issues related to the rights of women, the abolition of slavery, and supporting universal education. He was very positive about the onset of the French Revolution and became very politically involved in it, as evidenced by the mention of his name in *Les Misérables*. Mascart (2000) notes in the foreword of Borda's biography that Borda was not as avid as Condorcet in supporting the ideology of the Revolution "which earned him Condorcet's enmity." Condorcet was elected to represent Paris in the new Legislative Assembly, where he was chosen to serve in the position as Secretary. In 1792 he was involved with writing the declaration that called for the suspension of King Louis XVI and in the call for the National Convention, where he represented the Department of Aisne. Then, the very serious problems started when he became a member of the Committee on the Constitution, where he drafted a proposal for a new constitution that was consistent with the sentiments of the more moderate Girondin political group during the Revolution. This proposed constitution was then rejected in favor of the one that was supported by the more radical Jacobins group that was headed by Robespierre. Condorcet also showed independence by voting against the death penalty in the trial of Louis XVI. When it became known that Condorcet anonymously wrote a pamphlet to urge citizens to reject the constitution that was proposed by the Jacobins, he was accused of treason and was declared to be an enemy of the state to face execution. He went into hiding at the home of a friend, where he occupied his time by writing his most famous book *Sketch for a Historical Picture of the Progress of the Human Mind.*

When Condorcet later feared that he was going to be located by the authorities and thereby endanger his host, he attempted to escape. He was caught a few days later and was arrested after a suspicious innkeeper notified the authorities. Condorcet's deadly mistake was made when, as a mathematician-philosopher who had always lived as a gentleman of the period, he claimed to be a carpenter and asked the innkeeper to prepare an omelet made with 12 eggs. His appearance and demeanor made a highly implausible fit with his claim to be a traveling carpenter. He was subsequently arrested and imprisoned at Bourg-la-Reine, where he was found dead in his cell 2 days later on March 28, 1794. There are conflicting stories regarding the cause of Condorcet's death, some saying he died of exposure and some saying that he poisoned himself. Given the interaction between Borda and Condorcet that was outlined above, it is quite amazing that McLean (1995) notes a report that Borda risked his own life to make a plea for clemency on behalf of Condorcet while he was in hiding. Thorough biographies of Condorcet can be found in many sources, including Baker (1975), Badinter and Badinter (1988), and McLean and Hewitt (1994). Borda's fate during the French Revolution was much better than that of Condorcet in that while he was dismissed from his position in the Academy of Sciences during the height of the Reign of Terror, he was fully reinstated to his post in April 1795 and remained in that position until he died in 1799.

Given this background of the happenings that transpired during the historical period when Borda and Condorcet were doing their groundbreaking work and the interactions that took place between them, we now turn our attention to exactly what these two mathematicians had to say about elections and voting rules in the midst of their very tumultuous and rather fascinating lives.

1.3 The Historic Analyses of Borda and Condorcet

We begin by developing some basic definitions and concepts that are essential to understanding the work that Borda and Condorcet did. Their research can typically be explained most easily for the case of elections on three candidates $\{A, B, C\}$. Each voter will obviously cast their ballot based on their own particular preferences on the candidates, and we start to describe these individual preferences by letting $A \succ B$ denote the fact that a voter prefers Candidate A to B. A voter's preferences on the set of all pairs of candidates are *complete preferences* if there is a preference on each of the possible pairs. That is, each voter must have either $A \succ B$ or $B \succ A$ for all pairs of candidates like A and B. When an individual voter's preferences are complete, there is no indifference allowed between any two candidates. We initially assume that all voters have complete preferences on the candidates, and the effect of individual voter indifference between candidates will be considered later.

We also assume that each individual voter has *transitive preferences* on the candidates, which is a common requirement in defining rational behavior in the context of the preferences of individuals. Transitivity requires that if a given voter has $A \succ B$ and $B \succ C$, then that voter must also have $A \succ C$, to prevent situations in which a given voter might respond in a cyclic fashion, such as $A \succ B$, $B \succ C$ and $C \succ A$. Arguments for requiring individuals to have transitivity of preference typically rely on some form of the concept of being able to use such a voter as a 'money pump' if such cyclic preferences truly exist. To describe this phenomenon, we assume that A, B and C are items of different values that some subject would really like to possess and that the subject has far more than enough resources available to purchase any of them. Further suppose that this subject actually has the cyclic preferences with $A \succ B$, $B \succ C$ and $C \succ A$. The problems begin when somebody approaches the unwitting subject and makes the apparently generous offer to give them an A for free, which would be gratefully accepted since this currently happy subject wants to have an A. The subject is then given the option to make a very small payment to receive C as a gift instead of A, and this offer would be accepted since the subject has the preference $C \succ A$ and the requested payment is small. The subject now has C, and they are now asked if they would like to make another very small payment to receive B instead of C. The subject would agree since $B \succ C$. The subject now has B, and they are then asked if they would like to make a very small payment to have A instead of B. The subject would agree since $A \succ B$, and after these transactions the subject is back in possession of A again. So, the series of three payments only leads back to the original situation. If the subject truly has these specified intransitive preferences, this process could be repeated until the subject's resources are exhausted and the ultimately unhappy subject will be in possession of an item of far lesser value than the total amount that was paid for all of the transactions to upgrade from the original purportedly free A. Any individual who actually has intransitive preferences clearly does not appear to be a person who is thinking rationally.

Condorcet (1788a, p. 160) discusses the issue of individuals with intransitive preferences and refers to them as being "self-contradictory." His view of this phenomenon is made very obvious when he writes that outcomes of this nature are "absurdities", and that voting rules must be designed to prevent them from occurring. However, scenarios can be developed to justify the possible existence of intransitive preferences (see for example Gehrlein 1990).

Individual voter preferences on candidates that are complete and transitive are defined as *linear preference rankings*. The six possible linear preference rankings that each voter might have for three-candidate elections are shown in Fig. 1.1.

Here, n_i denotes the number of voters that have the associated linear preference ranking on the three candidates, so that n_1 voters all have individual preferences with $A \succ B \succ C$. Of course, transitivity requires that $A \succ C$ also. If we let n define the total number of voters, then $\sum_{i=1}^{6} n_i = n$. Any particular combination of n_i terms that sum to n is referred to as a *voting situation*, $\mathbf{n}$. Voting situations just list the n_i values that are associated with the possible preference rankings in a given election, without specifying the particular preferences of any individual voter. A *voter profile* consists of a list of each of the individual voter's identities along with their respective linear preference ranking on candidates. Voting situations are therefore anonymous with respect to the individual voter's preferences while voter profiles are not anonymous.

We are now able to use these definitions to formally consider two different ways of extending the basic notion of majority rule to the case of more than two candidates. The most obvious of these extensions is the one that people typically think of and it is widely known as *Plurality Rule* (PR). Just as with two candidates, each voter casts a vote for his or her most preferred candidate with PR, and the winner is the candidate who receives the greatest total number of votes. Let ***APB*** denote the event that A beats B with PR voting. Assuming that all voters will cast votes in agreement with their true preferences according to Fig. 1.1, A will be the winner by PR for any voting situation if both ***APB*** with $n_1+n_2>n_3+n_5$ and ***APC***

Fig. 1.1 The six possible linear preference rankings on three candidates

A	A	B	C	B	C
B	C	A	A	C	B
C	B	C	B	A	A
n_1	n_2	n_3	n_4	n_5	n_6

with $n_1 + n_2 > n_4 + n_6$. It will typically be assumed throughout this study that voters will always cast votes in accordance with their true preferences, and the case in which this is not necessarily true is considered in the final chapter.

Borda (1784) considers the second extension of majority rule to three-candidate elections, by applying the basic majority rule relation to pairs of candidates. Let *AMB* denote the event that *A* is the majority rule winner over *B* when only *A* and *B* are considered. By ignoring the relative position of *C* in the possible preference ranking for any of the individual voter's rankings in Fig. 1.1, we see that *AMB* if $n_1 + n_2 + n_4 > n_3 + n_5 + n_6$. If *AMB*, then *A* beats *B* by *Pairwise Majority Rule* (PMR). We use the same basic logic for other pairs of candidates and *AMC* if $n_1 + n_2 + n_3 > n_4 + n_5 + n_6$ and *BMC* if $n_1 + n_3 + n_5 > n_2 + n_4 + n_6$.

Both Borda (1784) and Condorcet (1785) refer to *A* as having "plurality support" if both *AMB* and *AMC*, and Borda specifically states that such a candidate "is really the favorite." So, Borda made this connection in his 1770 presentation to the French Academy of Sciences and Condorcet wrote about it in his 1785 book, and such a winner has somehow become widely known as the *Condorcet Winner* (CW). The authors of the current study have avoided this apparent contradiction in earlier writing by referring to such a candidate as the *Pairwise Majority Rule Winner*, but we now change and adopt the more commonly used phrase of Condorcet Winner. We also present the further results of Borda and Condorcet by using terminology that is in current usage in the field, since their original wording can lead to significant confusion. If both *AMC* and *BMC* for the three-candidate case, then *C* would be a terrible choice for selection as the winner of an election and it is called the *Condorcet Loser* (CL). These definitions are extended in the obvious fashion when more than three candidates are considered.

1.3.1 Borda's Paradox

Borda (1784) then makes a very interesting observation regarding the possibility that an unusual election outcome can occur after he developed the idea of PMR. The term *voting paradox* is used to describe any election outcome that is contrary to what one would expect to observe in a stable environment, and Borda's original example of the particular voting paradox that he identified is based on the voting situation in Fig. 1.2 for 21 voters.

The concern in this example is related to the outcome of the election when PR is used, versus the outcome when PMR is used. In using PR with the voting situation in Fig. 1.2, *APB* (8-7), *APC* (8-6) and *BPC* (7-6) to give a linear ranking with *APBPC*. A very different result is observed using PMR, with *BMA* (13-8), *CMA* (13-8) and *CMB* (13-8) to give a linear ranking *CMBMA*. PR and PMR reverse the rankings on the candidates with this particular voting situation, and Borda found this phenomenon to be quite disturbing; leading him to assert that PR should never be used. Based on the result from this example, the definition of a *Strict Borda Paradox* has been generalized to cover any voting rule to describe the occurrence of

Fig. 1.2 An example voting situation displaying Borda's Paradox from Borda (1784)

A	A	B	C
B	C	C	B
C	B	A	A
$n_1 = 1$	$n_2 = 7$	$n_5 = 7$	$n_6 = 6$

a voting situation for which the ranking of candidates by that voting rule reverses that candidate ranking that is obtained by PMR.

Borda was primarily concerned by the less stringent possibility that the CL would be chosen as the winner by PR. The definition of an occurrence of a *Strong Borda Paradox* with a given voting rule accordingly describes the case when a voting situation is observed for which that voting rule elects the CL, without necessarily having a complete reversal in the candidate rankings that are obtained with that voting rule and PMR.

1.3.2 Borda's Solution to the Possibility of Borda's Paradox

Borda (1784) proposed two different election procedures to deal with the possibility that the various forms of Borda's Paradox might occur. The first procedure simply calls for using PMR on all pairs of candidates to directly determine the CW. With the additional observation that using such a process would be extremely time-consuming, Borda makes a second suggestion to use Election by Order of Merit. This voting rule states that each voter should first completely rank all of the candidates, and then each voter's most preferred candidate in an m candidate election should receive m points, the second most preferred candidate should receive $m - 1$ points, ... , and the least preferred candidate should receive one point. The winner is determined by summing the points that are received by each candidate from all of the voters, and declaring the candidate with the most total points as the winner. The points that any voter awards to candidates with this scheme reduces to the rank that the candidate has in that voter's preference ranking on the candidates, where one refers to a voter's least preferred candidate and a rank of m refers to the voter's most-preferred candidate. Nicolas Cusanus suggested a voting rule exactly like Election by Order of Merit some 400 years before Borda's work appeared, but no mathematical justification was provided at that time to support using it (McLean 1990).

For any general voting situation with voters' preference rankings as described in Fig. 1.1 with n voters and three candidates, the points that are scored for A, B and C with Election by Order of Merit are respectively *Score*(A), *Score*(B) and *Score*(C) with:

$$Score(A) = 3(n_1 + n_2) + 2(n_3 + n_4) + 1(n_5 + n_6)$$
$$Score(B) = 3(n_3 + n_5) + 2(n_1 + n_6) + 1(n_2 + n_4)$$
$$Score(C) = 3(n_4 + n_6) + 2(n_2 + n_5) + 1(n_1 + n_3).$$

The particular example from Borda that is given in Fig. 1.2 results in $Score(C) = 47$, $Score(B) = 42$, and $Score(A) = 37$. If we let ***ABB*** denote the event that Candidate A beats B by this rule, Borda's example yields a linear ranking on the candidates with ***CBBBA***, which reverses the order of the PR ranking and as a result it is in perfect agreement with the ranking by PMR. For this example at least, Election by Order of Merit looks far superior to PR for selecting a winner. It was mentioned earlier that Borda presented his work to the French Academy of Sciences in 1770, and the Academy later adopted its use to elect new members. It was used by the Academy until 1801 when Napoleon Bonaparte insisted that another rule that he favored should replace it.

Some authors suggest that Borda was claiming that Election by Order of Merit would always select the CW, but he never actually made this assertion (see for example Nurmi 1999). As mentioned before, Borda was primarily concerned with the notion that the CL should never be elected. Daunou (1803) presented a proof that this voting rule cannot rank the CW in last place in a general m-candidate election. This result was later extended in Fishburn and Gehrlein (1976) to prove that Election by Order of Merit cannot select the CL as the unique winner in an m-candidate election, which is precisely the criterion that Borda was promoting.

The issue on intensity of preference is raised again in Borda's analysis in a different format than we considered it while discussing majority rule voting in two-candidate elections. If candidates are not actually equally spaced along the preference spectrum of a voter, then the linearly decreasing weights of this rule do not accurately reflect that voter's true relative degree of preferences between the candidates. For example, if the voter has a linear preference ranking with $A \succ B \succ C$, the use of the linearly decreasing 3:2:1 point scale results in the same difference in point values being assigned as we move from any candidate in the voter's preference ranking to the candidate ranked immediately below it. This inherently assumes that B is not considered as being closer to A in the voter's true preference spectrum than it is to C, and it is not considered to be closer to C than it is to A.

Borda (1784, p. 115) clearly does not support the notion of making any attempt to account for the intensity of any particular voter's preferences by allowing the use of any nonlinear system to assign scores to candidates, when he argues that all voters must be treated equally and have their preference rankings represented by the "same degree of merit" as any other voter. The famous mathematician Laplace (1795) addressed the use of linearly decreasing weights with Election by Order of Merit, and justifies the use of such weights from the perspective of expected values for general m-candidate elections.

Daunou (1803, pp. 262–263) does not agree with the analysis that is presented by Borda and Laplace for more than two candidates, since they start with model that allows voters to express their degrees of personal preferences for the candidates, and then proceed to require the use of expected values for all voters, to thereby deprive voters of that possibility to express the nuances of their preferences. As mentioned previously, most studies of election procedures ignore the issue of intensity of preference, since it greatly complicates matters with implementing elections to do so, and we generally follow that same direction until the final chapter.

1.3.3 Condorcet's Paradox

To begin our analysis of Condorcet's work, we note that a significant problem arises when doing this, due to the fact that his writing style and the lines of reasoning often make it very difficult to understand what he is explaining. Todhunter discusses the work of both Borda and Condorcet in *A History of the Mathematical Theory of Probability*, and makes an extremely harsh criticism to give some insight into the difficulty that can be encountered in following Condorcet's arguments (Todhunter 1865, p. 352):

> "Condorcet's work is excessively difficult; but the difficulty does not lie in the mathematical investigations, but in the expressions that are employed to introduce these investigations and to state their results: it is in many cases almost impossible to discover what Condorcet means to say. The obscurity and self-contradiction are without parallel. ... We believe that the work has been very little studied, for we have not observed any recognition of the repulsive peculiarities by which it is so undesirably distinguished."

Todhunter was not alone in his assessments of Condorcet's writing style. Baker (1975, p. 6) notes that Condorcet was indeed admitted to the French Academy of Sciences in 1769, but it is added that Condorcet had originally submitted his work to the Academy for their evaluation in 1761. The written evaluation of that work stated that it showed mathematical knowledge, but it was condemned for its lack of clarity and sloppiness, and it was returned to Condorcet. Given that background, we summarize the results from Condorcet's book that are relevant to the analysis that we are considering.

Condorcet's 1785 book extended some of Borda's ideas, and its tone becomes increasingly more antagonistic toward Borda's work as it progresses. This analysis begins with an example voting situation that leads to the outcome that the candidate ranking by PR is the reverse of the ranking by PMR, just as in the result given in Borda's example in Fig. 1.2. Condorcet then follows Borda's reasoning again by stating that this outcome results from ignoring additional information that could be obtained by requiring voters to report linear preference rankings on candidates "in order of merit", with no mention of Borda's earlier work.

Condorcet routinely stressed the concept that the winner of any election should be the CW. Since it is difficult to argue against this criterion, which has come to be known as the *Condorcet Criterion*, it can come as quite a surprise to find that a CW

A	B	C	B	C
B	A	A	C	B
C	C	B	A	A
$n_1 = 23$	$n_3 = 2$	$n_4 = 10$	$n_5 = 17$	$n_6 = 8$

Fig. 1.3 A voting situation with a PMR cycle from Condorcet (1785)

does not always exist. Condorcet proved this possibility with the famous example of a voting situation with 60 voters on three candidates in Fig. 1.3.

Condorcet describes the outcome from this voting situation as representing a "*contradictory system*" and it serves as an example of what has come to be widely known as *Condorcet's Paradox*. In particular, we find that PMR comparison leads to: *AMB* (33-27), *BMC* (42-18), and *CMA* (35-25). There is a cycle in this relationship on the three candidates, so that no candidate emerges as being superior to each of the remaining candidates. No matter which candidate we might choose as the winner of the election, a majority of the voters would prefer to have some other candidate to be the winner. Condorcet was very adamant in his argument that a lack of transitivity of preference for individual voters was so contradictory, that a system must be used to eliminate "such absurdities." However, after eliminating all intransitivity from the preferences of individual voters, we find that collective choice of voters from PMR still might produce intransitive results, suggesting an irrational response in the overall collective choice of otherwise rational voters. Given Condorcet's strong arguments that the CW should always be selected as the winning candidate, we are left with a very difficult question in this case: "Which candidate should be selected as the winner?"

Condorcet describes this outcome in mild terms as reflecting "some uncertainty of opinion" from the electorate. The number of papers that have been written about the possibility that Condorcet's Paradox might be observed is mind boggling, and the general consensus is that transitivity of collective choice must remain as a standard of rational behavior for group decisions. We paraphrase many of the negative connotations that have been attributed to a lack of transitivity as: lacking inner harmony, discordant, anarchic, democratically unpalatable, chaotic, leading to inconsistencies, politically incoherent, leading to arbitrary political decisions, lacking viability, leading to instability, and leading to pathology.

The existence of Condorcet's Paradox can be the source of great mischief when it appears in a voting situation. Strong concern about the possibility of manipulative behavior on the part of the chairperson of a committee was expressed by Riker (1958), who looked at actual voting events in the US Senate. When there are three alternative options for a specific bill that is being voted on, Senate rules require that the chairperson should select two alternatives, and majority rule voting will determine which of these two alternatives will move on to be compared to the remaining alternative by majority rule vote to determine the ultimate winner. Because of this policy, the chairperson can obtain any desired outcome when Condorcet's Paradox exists in a voting situation, by appropriately setting the order in which the alternatives are presented for sequential elimination. It can become much more manipulative than that if the chairperson does not like the outcome that would be obtained when only two real options are presented. A contrived third option could be introduced to the agenda to create a scenario that exhibits Condorcet's Paradox, and the chairperson could thereby select the order in which the sequential voting takes place to have the original losing real option become the final winner from the three options. Riker produces evidence to support the premise that this highly manipulative strategy has actually been practiced in the Senate.

One of the most famous works in the area of social choice theory was written by Kenneth Arrow, where a list of logical requirements for rational behavior for group decision making was developed. This list obviously includes transitivity of group choice. The amazing outcome of this work is a mathematical proof that no voting rule, other than having a dictator, can simultaneously satisfy all of these listed conditions. Arrow obviously was not arguing for the establishment of dictatorships, so normal situations demand that at least one of the requirements has to be abandoned. Arrow argues that requiring transitivity of group preference from voting procedures, including PMR, is an important aspect of describing rational behavior in collective choice. However, he acknowledges the work of other researchers in the area (Arrow 1963, p. 118) who suggest "that a social decision process might well sacrifice transitivity if necessary to satisfy other conditions." These other researchers are therefore identifying transitivity of group preference as the weakest of the requirements in that list (see for example Fishburn 1970).

Condorcet then continues with his analysis of the presence of intransitive PMR in voting situations, to show that there might be a CW with more than three candidates, while there is a cycle in the PMR relationships on the remaining candidates. A distinction is therefore being made between the possibility that there is a CW and the possibility that PMR is completely transitive over the entire set of candidates. When there are only three candidates, the existence of a CW ensures that this ranking is transitive over all candidates. Condorcet adds that the possible existence of PMR cycles with more than three candidates is of no consequence to the general superiority of the CW if there is one, as long as only one candidate is being elected as the winner.

The comments of Todhunter above suggested that Condorcet's 1785 book had been studied very little up to that point due to the difficulty that is encountered in trying to read it. The basic concept of Condorcet's Paradox was however rediscovered by several other people, starting a century later with Dodgson (1885). An interesting historic sidelight is that Dodgson is much better known by the penname Lewis Carrol which he used to write *Alice's Adventures in Wonderland*. Condorcet's Paradox was also rediscovered in a study by Huntingdon (1938), and more thorough reviews of the history of Condorcet's Paradox can be found in Granger (1956), Black (1958) and Riker (1961).

1.3.4 Condorcet's Other Paradox

Condorcet then goes on to develop the general notion of a Weighted Scoring Rule, which assigns some number of points to candidates according to their relative position within individual voter's preference rankings, just like Election by Order of Merit. In modern terminology, a Weighted Scoring Rule *WSR*(λ) for three candidates assigns one point to a candidate for each most-preferred ranking in a voter's preferences, λ points for each second-place ranking, and zero points for each least-preferred ranking. It would not make any logical sense to award more points to the middle-ranked candidate in a voter's preference ranking than we award to the most-preferred candidate in that ranking, or to award fewer points to the middle-ranked candidate than we award to the least-preferred candidate, so it is required that $0 \leq \lambda \leq 1$.The earlier definition of Borda's Election by Order of Merit corresponds exactly to the special case of a *WSR*(λ) with $\lambda = 1/2$, which we use to define Borda Rule (BR), and the case with $\lambda = 0$ corresponds to PR.

Condorcet uses this definition of a *WSR*(λ) to develop another example voting situation that illustrates the third voting paradox that we consider. This example voting situation with 81 voters is shown in Fig. 1.4.

The use of PMR with this voting situation results in *AMB* (41-40) and *AMC* (60-21), so that we have Candidate *A* as the CW; and when the *WSR*(λ) version of BR is used in this example

$$Score(A) = 1(31) + \frac{1}{2}(39) + 0(11) = 50.5$$

$$Score(B) = 1(39) + \frac{1}{2}(31) + 0(11) = 54.5.$$

So, we have *BBA* when *A* is the CW, to show again that BR does not always elect the CW. And, Condorcet then goes on to demonstrate *Condorcet's Other Paradox*. This argument involves using a general *WSR*(λ) with this voting situation in Fig. 1.4 to compute *Score*(*A*) and *Score*(*B*) for a general value of λ:

$$Score(A) = 1(31) + \lambda(39) + 0(11)$$
$$Score(B) = 1(39) + \lambda(31) + 0(11).$$

Since BR does not elect the CW in this example, Condorcet questions what value of λ would accomplish that outcome. This would require:

A	*A*	*B*	*C*	*B*	*C*
B	*C*	*A*	*A*	*C*	*B*
C	*B*	*C*	*B*	*A*	*A*
$n_1 = 30$	$n_2 = 1$	$n_3 = 29$	$n_4 = 10$	$n_5 = 10$	$n_6 = 1$

Fig. 1.4 A voting situation showing Condorcet's Other Paradox from Condorcet (1788b)

A	A	B	C	B	C
B	C	A	A	C	B
C	B	C	B	A	A
$n_1 = 9$	$n_2 = 3$	$n_3 = 4$	$n_4 = 4$	$n_5 = 6$	$n_6 = 4$

Fig. 1.5 A voting situation where PR elects the CW and BR does not from Condorcet (1788b)

$$\begin{aligned}
&Score(A) > Score(B)\\
&1(31) + \lambda(39) + 0(11) > 1(39) + \lambda(31) + 0(11)\\
&31 + 39\lambda > 39 + 31\lambda\\
&8\lambda > 8\\
&\lambda > 1.
\end{aligned}$$

The outcome $\lambda > 1$ completely contradicts our basic definition of a *WSR*(λ), and it then directly follows that no *WSR*(λ), including BR, can elect the CW in this example voting situation, which is Condorcet's Other Paradox.

Condorcet (1788b) considers Borda's work again and becomes more aggressive in his attack on BR, by presenting the example voting situation that is shown in Fig. 1.5 for 30 voters on three candidates.

With this particular example, we find *AMB* (16-14) and *AMC* (16-14), so Candidate A is the CW. PR gives Candidate A 12 votes, B 10 votes, and C 8 votes; so that PR elects the CW. And, the scores for candidates with BR are:

$$\begin{aligned}
Score(A) &= 1(12) + \frac{1}{2}(8) + 0(10) = 16\\
Score(B) &= 1(10) + \frac{1}{2}(13) + 0(7) = 16.5\\
Score(C) &= 1(8) + \frac{1}{2}(9) + 0(13) = 12.5.
\end{aligned}$$

So, BR fails to elect the CW for this example, since we have *BBA*, while the use of PR, which Borda thoroughly discredited, does elect Candidate A. This example leads to the strongly stated conclusion to condemn BR in Condorcet (1788b, p. 151) that asserts that while PR only suffers from the possibility that it might produce results that do not agree with the will of the electorate, it is certain that BR is worse and that it will fail to do so.

1.3.5 Borda Versus Condorcet: Over 200 Years Later

The written disagreements between the positions that were put forward by Borda and Condorcet themselves certainly ended when Condorcet died in prison in 1794,

Fig. 1.6 Example Voter Profile 1 from Saari (1995)

Voter 1: $A \succ B,\ B \succ C,\ A \succ C$
Voter 2: $B \succ C,\ C \succ A,\ B \succ A$
Voter 3: $C \succ A,\ A \succ B,\ C \succ B.$

Fig. 1.7 Example Voter Profile 2 from Saari (1995)

Voter 1: $A \succ B,\ B \succ C,\ C \succ A$
Voter 2: $A \succ B,\ B \succ C,\ C \succ A$
Voter 3: $B \succ A,\ A \succ C,\ C \succ B.$

and it might seem very reasonable to assume that some consensus has been reached from the enormous number of papers that have been written about their ideas since then. But, when dealing with these philosophically based arguments, such assumptions often do not hold.

The notion that the CW should be elected as the winner of an election has most certainly received significant support, to lead to the conclusion that methods should be sought to elect such candidates (see for example Felsenthal and Machover 1992). Because a CW does not always exist, Black (1958) suggests a compromise between the two positions with a hybrid model that will elect the CW whenever such a candidate exists, and then use BR when a CW does not exist.

The most outspoken critic of the basic Condorcet Criterion is Donald Saari, who strongly supports the direct use of BR in all elections. To describe Saari's opposition to using any form of PMR to determine the winner in an election, two different voter profiles are defined for a three-candidate election. Voter Profile 1 is shown in Fig. 1.6, and it lists the preference rankings on candidates for three rational voters.

By using PMR with the preference rankings in Voter Profile 1, an example of Condorcet's Paradox results, with *AMB* (2-1), *BMC* (2-1) and *CMA* (2-1).

Voter Profile 2 in Fig. 1.7 illustrates another scenario in which there are three irrational voters who all have intransitive preferences on the candidates.

The outcome *AMB* (2-1), *BMC* (2-1) and *CMA* (2-1) from using PMR on Voter Profile 2 turns out to be identical to the results that were obtained from Voter Profile 1. Saari (1995, p. 48) then describes the observation that identical results are obtained from both the rational voters in Voter Profile 1 and the irrational voters in Voter Profile 2 as a fatal flaw for the use of PMR. It is noted that Condorcet was very careful to require transitivity as a condition of rationality for the individual voters, but then proceeded to suggest the use of PMR which "surreptitiously drops it" as a necessary condition for the aggregated behavior for the electorate. However, Saari (1995, p. 46) does acknowledge that the Condorcet Criterion does indeed have "nearly universal acceptance."

The net result of all of this discussion is that the Condorcet Criterion is definitely alive and doing very well. But, it must also be stressed very strongly that there is absolutely no reason to dismiss BR, as Condorcet suggested, since it will be found to have many interesting properties. It turns out that, in their own way, Borda and Condorcet were both basically correct in their arguments. Before we go on to pursue that line of reasoning, we look at some of the many other interesting voting paradoxes that have been identified since Borda and Condorcet did their very important groundbreaking work.

1.4 Other Voting Paradoxes

Many voting paradoxes have been identified in the literature, and they have been categorized into groups of paradox types that have a similar basic theme. We look primarily at two of these categories that are called Monotonicity Paradoxes and Choice Set Variance Paradoxes, but it is necessary to introduce a few additional voting rules before some of them can be explained. To begin, *Negative Plurality Rule* (NPR) has voters cast a vote for each of their two more-preferred candidates in a three-candidate election, and the candidate that receives the most votes is the winner. So, NPR is identical to a *WSR*(λ) with $\lambda = 1$, and it is also referred to as *Antiplurality Rule* since it is equivalent to having each voter cast a vote against their least-preferred candidate, where the winner is the candidate that receives the smallest number of these negative votes.

Elimination procedures are two-stage election processes for three-candidate elections. Some candidate is chosen for elimination in the first stage, and majority rule is then used to determine the ultimate winner from the two remaining candidates that are carried on to the second stage of the election. The only difference is the rule that is used to identify the candidate for elimination in the first stage. *Plurality Elimination Rule* (PER) eliminates the candidate that receives the fewest votes by PR, and *Negative Plurality Elimination Rule* (NPER) eliminates the candidate that receives the fewest votes by NPR.

1.4.1 Monotonicity Paradoxes

Monotonicity Paradoxes represent situations in which some reasonable definition has been established to determine which candidate should be viewed as being the 'best' available candidate based on the voters' preferences, and where the voting rule that has been selected for use to determine the winner does not behave in a monotonic fashion. *Monotonicity* of a voting rule basically requires that there is a consistent relationship between the determination of an election outcome and any changes that occur in voters' preferences. To be more specific, increased support for a given candidate among the voters' preferences should not prove to be detrimental towards that candidate's chances for winning the election, and reduced support for a candidate in voters' preferences should not improve the standing of that candidate in election results.

The *No Show Paradox* is developed in Brams and Fishburn (1983), with an example voting situation for which some subset of voters chooses not to participate in an election, and where they prefer the resulting winner to the candidate that

A	A	B	C	B	C
B	C	A	A	C	B
C	B	C	B	A	A
$n_1 = 5$	$n_2 = 7$	$n_3 = 7$	$n_4 = 4$	$n_5 = 5$	$n_6 = 5$

Fig. 1.8 An example voting situation for the No Show Paradox

would have been selected if they had participated. As an example of this paradox, we start with the voting situation for 33 voters that is shown in Fig. 1.8, where the winner of an election is determined by NPER.

NPR is used in the first stage of voting, and Candidates A, B, and C will receive 23, 22 and 21 votes respectively. Candidate C receives the fewest votes, so it is therefore eliminated in the first stage of voting. The two surviving Candidates A and B then go on to the second stage where majority rule is used to find BMA by a vote of 17-16 to select B as the overall election winner.

Suppose that two good friends who participated in this particular election were talking with each other after B had been declared the winner, and each of them knew that they both had the same linear preference ranking on candidates with $A \succ B \succ C$. Neither of them would be at all happy that their most favored Candidate A had lost, but they do take consolation from the fact that their least favored Candidate C was not the winner, and that C was actually made to look very bad to the public by being quickly eliminated in the first round of voting. While the friends really wanted Candidate A to win, they also take some solace from the fact that their preferences had been accounted for as a result of their participation in the election, and that they thereby helped to promote their best interests to some degree. But, is all of this reasoning that leads to their limited level of contentment really valid?

Consider the scenario in which both of these friends had chosen to remain at home to watch political coverage of this election on television instead, so that they would not have voted in the election at all. The original voting situation from Fig. 1.8 then would have looked like the modified voting situation with only 31 participating voters in Fig. 1.9.

The first stage of voting with NPR in this modified voting situation gives Candidates A, B, and C respectively 21, 20 and 21 votes. Candidate B receives the fewest votes and it is therefore eliminated from further consideration in the first stage. Candidates A and C then move on to the second stage of voting where AMC by a vote of 17-14, so that Candidate A would have been declared the winner of this election if the two friends with the preference ranking $A \succ B \succ C$ had not participated in the election. The two friends therefore actually ended up with a worse outcome from their personal perspectives because they did choose to go and vote. Their favorite candidate was penalized as a result of the fact that their votes indicated increased support for that candidate, which is a clear violation of monotonicity. This outcome would obviously be quite disconcerting to the two friends and totally eliminate the solace that they found when they thought that they had made a positive impact on their own interests by taking the effort to go and vote.

A	A	B	C	B	C
B	C	A	A	C	B
C	B	C	B	A	A
$n_1 = 3$	$n_2 = 7$	$n_3 = 7$	$n_4 = 4$	$n_5 = 5$	$n_6 = 5$

Fig. 1.9 The modified example voting situation for the No Show Paradox

Occurrences of various Monotonicity Paradoxes are very often associated with the existence of Condorcet's Paradox, and the possibility of their occurrence is not unique to NPER.

1.4.2 *Choice Set Variance Paradoxes*

Choice Set Variance Paradoxes typically represent situations in which a series of m different issues are being presented to a committee or group of voters. Each of the individual issues that are under consideration will be independently approved or disapproved with the use of majority rule voting. A paradoxical voting outcome can then arise if the overall final approve and disapprove outcomes on the issues represents a general result that is in some manner inconsistent with the overall underlying preferences that the voters have on the issues. We consider three such Choice Set Variance Paradoxes in the form of Ostrogorski's Paradox, the Majority Paradox and the Referendum Paradox.

1.4.2.1 Ostrogorski's Paradox

We describe Ostrogorski's Paradox in the context of a scenario in which there are m individual issues that are being considered (Ostrogorski 1902). There are two competing political parties that are called Party R and Party L, that hold opposing positions on each of these issues. So, one of the two parties will take the position that any given issue should be approved, while the other party will prefer to have that particular issue disapproved. Each of the n individual voters has their own personal preference on each of the different issues that is in agreement with the position of either Party R or Party L. Some voters will likely be staunch supporters of one of the parties and have preferences that are completely aligned with the position of either Party R or Party L over all of the issues. But, there will likely also be some independent voters with preferences on issues that are not in agreement with the position of the same party on every issue.

Voters are then categorized as being affiliated with either Party R or Party L based on the level of agreement between their individual preferences on issues and the party positions on issues. For example, a voter is considered to be affiliated with Party R if their individual preferences are in agreement with Party R positions over a majority of the issues that are being considered. Otherwise, they would be

Issue	Voter					Position Winner
	1	2	3	4	5	
1	*R*	*R*	*R*	*L*	*L*	*R*
2	*R*	*R*	*L*	*L*	*R*	*R*
3	*R*	*R*	*L*	*R*	*L*	*R*
Party Affiliation	*R*	*R*	*L*	*L*	*L*	

Fig. 1.10 An example voting scenario that results in a Strict Ostrogorski Paradox

affiliated with Party L. Consider the example in Fig. 1.10 where each of five voters has preferences on three different issues.

The results in Fig. 1.10 show for example that Voters 1 and 2 are staunch supporters of Party *R* and have preferences on issues that completely agree with the position that is taken by that party on all three issues. Voter 4 is more independent and has preferences on Issues 1 and 2 that are in agreement with Party *L*, while being in agreement with Party *R* on Issue 3. Since Voter 4 is in agreement with Party *L* on a majority of issues by a 2-1 margin, this voter is listed as having an affiliation with Party *L*. Using this same logic, three of the five voters have an affiliation with Party *L*, to make it the *Majority Party* by a 3-2 margin.

The paradoxical outcome now occurs when we compare the fact that Party *L* is the Majority Party to the majority rule election outcomes on issues. When Issue 1 is voted on, Voters 1, 2 and 3 vote for the issue position that is taken by Party *R*, while Voters 4 and 5 agree with the Party *L* position. This yields an outcome for the Issue 1 election that is in agreement with the position of Party *R* by a 3-2 margin. That single outcome in itself does not seem to be terribly paradoxical, but we find that exactly the same thing also happens with both of the other issues. So, Party *L* is the Majority Party, but Party *R* somehow obtains the election outcome that it favors on every issue. This outcome has come to be known as a *Strict Ostrogorski Paradox*, and the possibility that such a contrary outcome might exist was first discussed over a century ago in Ostrogorski (1902). A less restrictive condition is a *Weak Ostrogorski Paradox* that occurs when either of Party *R* or Party *L* is the Majority Party, while a majority of the election outcomes on issues are in agreement with the issue positions that are held by the opposing party.

1.4.2.2 Majority Paradox

The *Majority Paradox* is similar in nature to Ostrogorski's Paradox, but it takes a different perspective to define its version of the Majority Party. No determination of the party affiliation of voters is performed to observe the Majority Paradox, and we are concerned instead about the number of majority rule outcomes on issues that are in agreement with the *Overall Majority Party*. For example, Party *R* is the Overall Majority Party if there are more Party *R* agreements than Party *L* agreements in the $(m)(n)$ different party position associations for preferences of the n voters over all of

Issue	Voter					Position Winner
	1	2	3	4	5	
1	*L*	*L*	*R*	*L*	*L*	*L*
2	*R*	*L*	*L*	*R*	*R*	*R*
3	*R*	*L*	*L*	*R*	*R*	*R*

Fig. 1.11 An example voting scenario that results in a Majority Paradox

the *m* issues. An example election outcome that illustrates the Majority Paradox is presented in Fig. 1.11.

This example shows the 15 different preference agreements that the five voters have with the parties on the three issues, with seven Party *R* agreements and eight Party *L* agreements. Since Party *L* agreement beats Party *R* agreement with majority rule by a margin of 8-7, Party *L* is the Overall Majority Party in this example. The determination of which party is the position winner for each of the issues is performed by majority rule, exactly following the logic that was used above for illustrating Ostrogorski's Paradox. The contrary outcome from the example in Fig. 1.11 is that Party *L* is the Overall Majority Party, while its opposing Party *R* is the position winner for a majority of the election outcomes on the issues. It is not possible to observe a strict version of the Majority Paradox, since any party that is the position winner by majority rule for every issue must obviously also be the Overall Majority Party.

1.4.2.3 Referendum Paradox

A *Referendum Paradox* is directly related to the Majority Paradox, and it is typically defined as occurring for indirect elections with two parties when one party gets a majority of the seats in a parliament without obtaining a majority of the votes nationwide. To illustrate this paradox in a slightly different format, consider a country with three election districts that each have 100 voters in Fig. 1.12. Elections are held between Parties *L* and *R* in each district, with a winner-take-all system determining the district winners

District 1 voters prefer Party *L* to Party *R* by 65 to 35 to make *L* the District 1 winner for the example in Fig. 1.12. By following the same process for the other two districts, Party *L* is the winner in two of the three districts so it would be declared to be the overall winner. However, Party *R* beats Party *L* by a majority of 165 to 135 in the vote count of the entire electorate, while it is beaten by Party *L*, which is the CL of the two candidates, in a majority of the districts.

Condorcet (1788a, pp. 167–168) presents an example of the Referendum Paradox, and further discusses the distinctions between the overall winner and the majority winner for the entire electorate. Miller (2012) notes that outcomes of the Referendum Paradox have been referred to as reversal of winners, unpopular winner, wrong winner, divided verdict, misfire, representative inconsistency, compound majority paradox, and majority deficit. Some actual occurrences of the

District	Preferred Candidate		District Winner
	L	R	
1	65	35	L
2	15	85	R
3	55	45	L
Total Votes	135	165	

Fig. 1.12 An example voting scenario that results in a Referendum Paradox

Referendum Paradox have received significant attention, with the most recent case being the election of Donald Trump over Hillary Clinton in the 2016 US Presidential election. With additional results from Neubauer et al. (2012), the Referendum Paradox has occurred at least five times for the 58 US Presidential elections that have taken place since 1789.

1.5 Conclusion

The study of elections and voting paradoxes has a very long history, and the mathematically based analyses on these topics by Borda and Condorcet laid the foundation of an enormous amount of work that has followed from their studies. Several voting rules have been introduced and it has been observed that most of them can be criticized from the standpoint that they can exhibit potentially serious issues. A number of voting paradoxes have been introduced to indicate that there are a number of disconcerting election outcomes that can occur, and many other paradoxes have been considered in sources such as Nurmi (1999).

The primary question that has to be considered at this point is whether or not all of these observations should lead us generally to conclude that election procedures are seriously flawed processes that lead to many types of unexpected and paradoxical outcomes. The other possibility is that these observations only represent some theoretical examples of unexpected outcomes that possibly could be observed, while there is only a very small likelihood that any of them will ever actually be seen in any real elections. This question can only be answered by addressing the issue of the probability that these paradoxical outcomes will be observed. It is not surprising to find that the work in this area finds its roots back in the studies that were presented by Borda and Condorcet. This approach was initiated by Condorcet (1793, p. 96) when he noted that "after considering the facts, the average values or the results, we still need to determine their probability."

The next chapter will introduce many of the basic concepts from probability analysis that have been introduced to the investigation of the likelihood that these paradoxical election outcomes will ever be observed in practice.

References

Arrow, K. J. (1963). *Social choice and individual values* (2nd ed.). New Haven CT: Yale University Press.

Badinter, E., & Badinter, R. (1988). *Condorcet: un intellectuel en politique*. Paris: Librarie Arthème Fayard.

Baker, K. M. (1975). *Condorcet: From natural philosophy to social mathematics*. Chicago IL: University of Chicago Press.

Black, D. (1958). *The theory of committees and elections*. Cambridge: Cambridge University Press.

Brams, S. J., & Fishburn, P. C. (1983). Paradoxes of preferential voting. *Mathematics Magazine, 56*, 207–214.

Daunou, P. C. F. (1803/1991). A paper on elections by ballot. In: F. Sommerlad, & I. McLean (Eds.), *The political theory of Condorcet II* (pp. 235–279). Oxford: University of Oxford Working Paper.

de Borda, J. C. (1784). A paper on elections by ballot. In I. McLean & F. Hewitt (Eds.), *Condorcet: Foundations of social choice and political theory* (pp. 114–119). Hants: Edward Elgar.

de Condorcet, M. (1785). An essay on the application of probability theory to plurality decision making. In I. McLean & F. Hewitt (Eds.), *Condorcet: Foundations of social choice and political theory* (pp. 120–130). Hants: Edward Elgar.

de Condorcet, M. (1788a). On the form of decisions made by plurality vote. In I. McLean & F. Hewitt (Eds.), *Condorcet: Foundations of social choice and political theory* (pp. 157–168). Hants: Edward Elgar.

de Condorcet, M. (1788b). On discovering the plurality will in an election. In I. McLean & F. Hewitt (Eds.), *Condorcet: Foundations of social choice and political theory* (pp. 148–156). Hants: Edward Elgar.

de Condorcet, M. (1793). A general survey of science - Concerning the application of calculus to the political and moral sciences. In I. McLean & F. Hewitt (Eds.), *Condorcet: Foundations of social choice and political theory* (pp. 93–98). Hants: Edward Elgar.

de Laplace, P. S. (1795). Analytic theory of probabilities. In: F. Sommerlad, I. McLean (1991, Eds.) *The political theory of Condorcet II* (pp. 282–286). Oxford: University of Oxford Working Paper.

Dodgson, C. (1885). *The principles of parliamentary representation: Postscript to supplement, E.* Oxford: Baxter Publisher.

Felsenthal, D. S., & Machover, M. (1992). After two centuries, should Condorcet's voting procedure be implemented? *Behavioral Science, 37*, 250–274.

Fishburn, P. C. (1970). The irrationality of transitivity of social choice. *Behavioral Science, 15*, 119–123.

Fishburn, P. C., & Gehrlein, W. V. (1976). Borda's rule, positional voting, and Condorcet's simple majority principle. *Public Choice, 28*, 79–88.

Gaertner, W. (2005). De jure naturae et gentium: Samuel von Pufendorf's contribution to social choice theory and economics. *Social Choice and Welfare, 25*, 231–241.

Gehrlein, W. V. (1990). Special issue on intransitive preferences. *Annals of Operations Research, 23*, 235–246.

Gehrlein, W. V. (2006). *Condorcet's paradox*. Berlin: Springer.

Granger, G. G. (1956). *La mathématique sociale du Marquis de Condorcet*. Paris: Presses Universitaires de France.

Huntington, E. V. (1938). A paradox in the scoring of competing teams. *Science, 8*, 287–288.

Lagerspetz, E. (1986). Pufendorf on collective decision. *Public Choice, 49*, 179–182.

Mascart, J. (2000). *La vie et les travaux du chevalier Jean-Charles de Borda (1733–1799), Episodes de la vie scientifique au XVIIIe siècle*. Paris: University of Paris-Sorbonne Press.

McLean, I. (1990). The Borda and Condorcet principles: Three medieval applications. *Social Choice and Welfare, 7*, 99–108.

McLean, I. (1995). The first golden age of social choice, 1784–1803. In W. Barnett, H. Moulin, M. Salles, & N. Schofield (Eds.), *Social choice, welfare, and ethics* (pp. 13–36). Cambridge: Cambridge University Press.

McLean, I., & Hewitt, F. (1994). *Condorcet: Foundations of social choice and political theory*. Hants: Edward Elgar.

Miller, N. (2012). Election inversions by the U.S. Electoral College. In D. S. Felsenthal & M. Machover (Eds.), *Electoral systems: Paradoxes, assumptions and procedures* (pp. 93–128). Berlin: Springer.

Neubauer, M. G., Schilling, M., & Zeitlin, J. (2012). *Exploring unpopular presidential elections*. Working paper, California State University.

Nurmi, H. (1999). *Voting paradoxes and how to deal with them*. Berlin: Springer.

Ostrogorski, M. (1902). *La démocratie et l'organisation des partis politiques*. Paris: Calmann-Levy Publishing.

Riker, W. H. (1958). The paradox of voting and congressional rules for voting on amendments. *American Political Science Review, 52*, 349–366.

Riker, W. H. (1961). Voting and the summation of preferences: An interpretive bibliographical review of selected developments during the last decade. *American Political Science Review, 55*, 900–911.

Rousseau, J. (1762). *The social contract, translated and reprinted in 1962 by Penguin Press*. England: Harmondsworth.

Saari, D. G. (1995). *Basic geometry of voting*. Berlin: Springer.

Todhunter, I. (1865). *A history of the mathematical theory of probability*. Cambridge: Macmillan.

Wilbour, C. E. (1909). *Les Misérables, Vol. 1* (Translated from the French). London: JM Dent and Son.

Young, P. (1995). Optimal voting rules. *The Journal of Economic Perspectives, 9*, 51–64.

McLean, I. [illegible]. The first golden age of social choice [illegible]. In [illegible] M. Salles & N. Schofield (Eds.), *Social choice, welfare, and ethics* (pp. [illegible]). Cambridge: Cambridge University Press.

McLean, I., & [illegible]

Miller, N. [illegible]. Election inversions by the U.S. Electoral College. In D. S. Felsenthal & M. Machover (Eds.), [illegible]

[illegible]

Chapter 2
Probabilities of Voting Paradoxes

2.1 Introduction

The logic of the general procedure that is used to determine representations for the probability that any given voting paradox might be observed is quite simple, and it is explained by considering the specific example of Condorcet's Paradox. We begin by replicating the six possible linear preference orders from the previous chapter that voters might have for the three candidates $\{A, B, C\}$ in Fig. 2.1.

As before, n_i represents the number of voters that have the associated preference ranking in a given voting situation. The added p_i terms in Fig. 2.1 will be used in further discussion to represent the probability that a randomly selected voter from the electorate will have the associated linear preference ranking. We used $\boldsymbol{n}$ to define any given voting situation with $\sum_{i=1}^{6} n_i = n$, and we now let $\boldsymbol{p}$ define any particular combination of p_i terms with $\sum_{i=1}^{6} p_i = 1$. The distribution of the p_i terms in $\boldsymbol{p}$ clearly defines the overall preference structure of the electorate.

There are two possible PMR cycles that can produce Condorcet's Paradox with three candidates. One has *AMB*, *BMC* and *CMA* and the other is the reverse cycle with *AMC*, *CMB* and *BMA*. The first cycle occurs when:

$$n_1 + n_2 + n_4 > n_3 + n_5 + n_6 \qquad [AMB]$$
$$n_1 + n_3 + n_5 > n_2 + n_4 + n_6 \qquad [BMC]$$
$$n_4 + n_5 + n_6 > n_1 + n_2 + n_3 \qquad [CMA].$$

All that we have to do in principle is to enumerate every possible voting situation that can exist for a specified n and identify the subset of them that meets the conditions of these three inequalities. The probability of observing such a cycle with *AMB*, *BMC* and *CMA* would then be obtained by summing the probabilities for observing the individual voting situations that are in that subset. The same process would then be repeated for the reverse cycle, and the probability that

W.V. Gehrlein, D. Lepelley, *Elections, Voting Rules and Paradoxical Outcomes*,
Studies in Choice and Welfare, DOI 10.1007/978-3-319-64659-6_2

Fig. 2.1 The six possible linear preference rankings on three candidates

A	A	B	C	B	C
B	C	A	A	C	B
C	B	C	B	A	A
n_1, p_1	n_2, p_2	n_3, p_3	n_4, p_4	n_5, p_5	n_6, p_6

Condorcet's Paradox is observed would simply follow as the sum of the probabilities for the two possible cycles. The resulting probability for observing Condorcet's Paradox will obviously be completely driven by the specific process that assigns a probability to observing each of the possible voting situations.

Four basic probability models have served as the primary foundation for early theoretical studies that assigned probabilities to voting situations to evaluate the likelihood that paradoxical outcomes will be observed: the Dual Culture Condition (DC), the Impartial Culture Condition (IC), the Uniform Culture Condition (UC) and the Impartial Anonymous Culture Condition (IAC). Each of these models will be described as we proceed, and subtle differences between these models will be pointed out as they are considered. These differences actually have a significant impact on the characteristics of the voting situations that are obtained from each of the models. We start with the DC and IC models, which both assume that each voter forms an individual linear preference ranking on candidates independently, without being influenced by the preferences of any other voter.

2.2 Models with Independent Voter Preferences

The basic concept of these models is most easily understood in the context of how they can be used to obtain a randomly generated voter profile from an electorate. Following the traditional methods that are used in any analysis of this type of probability modeling, we start with an urn that contains some total number of balls, with each ball being one of six different colors. Each of these colors corresponds to one of the six possible linear preference rankings on the three candidates in Fig. 2.1. The actual total number of balls that are contained in the urn is of no particular consequence at this time, but the proportional mix of the colors is critical. The proportions of the number of balls of each color must be equal to the probabilities for observing the associated linear preference ranking on candidates for some specified $\boldsymbol{p}$ in Fig. 2.1. So, if $p_3 = .2$ for $\boldsymbol{p}$ then 20% of the balls that are contained in the urn must be of the color that is associated with the linear preference ranking with $B \succ A \succ C$.

Balls are sequentially drawn at random from the urn over n different trials to assign linear preference rankings to each of the voters. The color of the selected ball on the ith trial will be used to assign the associated linear preference ranking on candidates to the ith voter, with the ball being returned to the urn after its color is noted. The random selection of balls is being done with replacement during the experiment so that the probability of observing any particular possible preference ranking for an individual voter does not change from draw to draw.

The voting situation $\boldsymbol{n}$ that results from any given voter profile that we obtain with such an experiment is obtained simply by determining the number of voters that have each of the six possible linear preference rankings. The factor of primary importance is that the probability that a voter is assigned any given linear preference ranking on candidates is determined solely by the proportions in $\boldsymbol{p}$, and this never changes during the experiment, so the preferences of the voters are statistically independent.

Let $\Delta(A,B)$ denote the difference between the sum of the p_i values for preference rankings with $A \succ B$ and $B \succ A$. The same definition is extended in the obvious fashion to all pairs of candidates, so that:

$$\begin{aligned}
\Delta(A,B) &= p_1 + p_2 + p_4 - p_3 - p_5 - p_6 \\
\Delta(A,C) &= p_1 + p_2 + p_3 - p_4 - p_5 - p_6 \\
\Delta(B,C) &= p_1 + p_3 + p_5 - p_2 - p_4 - p_6.
\end{aligned}$$

When independence exists between the voter's preferences, a randomly selected voter will be more likely to have a preference ranking with $A \succ B$ than with $B \succ A$ from these definitions whenever $\Delta(A,B) > 0$. The Law of Large Numbers will apply when there is a large electorate with $n \to \infty$, and we will therefore have *AMB* with a probability approaching one for any corresponding voting situation that is drawn from a specified $\boldsymbol{p}$ whenever $\Delta(A,B) > 0$.

As a result, Candidate A will be the CW whenever both $\Delta(A,B) > 0$ and $\Delta(A,C) > 0$, B will be the CW if both $\Delta(A,B) < 0$ and $\Delta(B,C) > 0$, and C will be the CW if both $\Delta(A,C) < 0$ and $\Delta(B,C) < 0$. The PMR cycle *AMB*, *BMC* and *CMA* will exist if $\Delta(A,B) > 0$, $\Delta(B,C) > 0$ and $\Delta(A,C) < 0$, and the reverse cycle *AMC*, *CMB* and *BMA* will exist if each of $\Delta(A,C) > 0$, $\Delta(B,C) < 0$ and $\Delta(A,B) < 0$. So, for any $\boldsymbol{p}$ as $n \to \infty$, the urn experiment described above will either have a CW with near certainty, or exhibit Condorcet's Paradox with near certainty, except for one particular class of $\boldsymbol{p}$ vectors.

2.2.1 The Dual Culture Condition

The DC assumption requires that a very specific format exists for $\boldsymbol{p}$ vectors such that the p_i terms yield an expected perfect balance of preferences on all pairs of candidates in the preference ranking of any randomly selected voter, with $\Delta(A,B) = \Delta(A,C) = \Delta(B,C) = 0$. A randomly selected voter is therefore equally likely to have $A \succ B$ or $B \succ A$, and the same is true for all pairs of candidates. Any $\boldsymbol{p}$ that exhibits this knife-edged perfect balance with DC will neither intentionally force a CW to exist nor intentionally force Condorcet's Paradox to occur in any associated randomly generated voting situation. This expected perfect balance in individual voter's preferences only exists with $\boldsymbol{p}$ vectors that meet the restriction of DC, which specifies $p_1 = p_6$, $p_2 = p_5$ and $p_3 = p_4$. By looking at Fig. 2.1, this definition requires that the probability that a randomly selected voter will have any particular preference ranking on the candidates is identical to the probability that this voter will have the dual, or inverted, preference ranking on the candidates.

Any results that are obtained regarding the probability that paradoxical voting outcomes are observed while using the DC model must be viewed in the context of what this assumption is saying about voter's preferences. In particular, any results that are obtained with DC represent a rather extreme case in which no candidate has any expected advantage whatsoever when the preferences on pairs of candidates are examined for any randomly selected voter. It is very important to emphasize that this balance of preferences with DC applies to individual voter's preferences on pairs of candidates. It does not preclude the possibility that some candidates might be ranked as most preferred, or least preferred, with greater likelihood than some other candidate in the preference ranking of a randomly selected voter. For example, DC applies to the case with $p_1 = p_6 = \frac{1}{2} - 2\varepsilon$ and $p_2 = p_3 = p_4 = p_5 = \varepsilon$ for small $\varepsilon > 0$, so that both Candidates A and C will be very likely to be either most preferred, or least preferred, in the voter preference rankings. Candidate B would therefore be very likely to be the middle-ranked candidate in voter preferences.

Sen (1970) presents a scenario like this when he considers a "two-class society" where the classes have radically different interests. For this "class war" type of situation, we could expect to have voter profiles that contain only two different rankings on alternatives. One class would have some preference ranking and the other class would have the dual preference ranking. If the two societies in this example were about the same size in number, this would be very much in agreement with the DC model in the immediately preceding discussion with $\varepsilon = 0$. Sen also notes that a CW must exist under this scenario for any odd number of voters. It is most assuredly not suggested that such scenarios would ever reflect a desirable outcome in society, but they really might exist, and DC could be used as a basis to model the probability that voting paradoxes might be observed as the preferences of the electorate approach such a "two-class" extreme as ε decreases to zero.

Another observation that follows from the analysis of the assumption of perfectly balanced preferences comes from a consideration of the resulting proportions of voters with preferences on pairs of candidates in a voting situation. That is, the proportion of voters with $A \succ B$ in a random voting situation will approach one-half with certainty as $n \to \infty$ if $\Delta(A, B) = 0$. The relative margins of all PMR wins and losses on pairs of candidates in voting situations will therefore be relatively small with perfectly balanced preferences for individual voters. As a result, this will lead to an environment that is conducive to the occurrence of voting paradoxes that involve PMR cycles in voting situations. When the assumption of DC is being utilized, we can therefore expect that exaggerated estimates will be obtained for the likelihood that voting paradoxes that involve PMR cycles will be observed in the resulting voting situations. But, it is important to stress that the DC assumption is neither forcing a CW to exist nor is it forcing Condorcet's Paradox to occur as $n \rightarrow \infty$.

2.2.2 *Impartial Culture Condition*

A refinement of DC is IC, which further assumes that each of the $m!$ possible preference rankings on the candidates in an m-candidate election is equally likely to represent the preferences of a randomly selected voter, so $p_i = \frac{1}{m!}$ for $1 \leq i \leq m!$. Since IC is a special case of DC, the preferences of any given voter are still independent of all other voters' preferences, and the expectation of perfectly balanced preferences holds on all pairs of candidates for a randomly selected voter. The additional restriction of IC beyond DC requires that there is also a perfect balance for the expected ranking position of all candidates, so that all candidates are equally likely to be most preferred, least preferred or middle ranked for a randomly selected voter. All of these assumptions make IC the 'purest' assumption, since no candidate will have any advantage whatsoever when it is compared to any other candidates in the preference rankings of a randomly selected voter.

No assertion has ever been made by anyone to suggest that IC could possibly ever be expected to reflect the preferences of voters in actual electorates. It does however serve as an ideal benchmark to observe what happens to the probability that voting paradoxes might occur, as characteristics of voters' preferences are changed to move away from this purest possible assumption regarding possible $\boldsymbol{p}$ that might be used to define the preferences of the electorate.

2.2.3 *Condorcet's Paradox and Social Homogeneity with DC*

Social homogeneity describes the degree to which voters' preferences tend to be in agreement, and it sounds very reasonable to say that paradoxical outcomes like Condorcet's Paradox should tend to become less likely to occur as voters' preferences on candidates become more alike. When all voters have exactly the same preference ranking on candidates, it is very difficult to think that any reasonable voting procedure would fail to select the most preferred candidate of every voter as the winner of an election. At the opposite extreme we have a scenario like IC where voters almost appear to be determining their preferences on candidates by rolling dice, with no inherent logical structure to the preference formation process. That is, no candidate is expected to have any advantage when compared to any other candidate, and every candidate is expected to be ranked as most preferred, middle-ranked and least preferred with equal likelihood. This describes an electorate for which the concept of social homogeneity appears to be quite meaningless under normal conditions.

Many different metrics have been proposed to measure the degree of social homogeneity that is reflected by an electorate with a given $\boldsymbol{p}$ to describe its preferences. The simplest of these measures is $H(\boldsymbol{p})$, which is defined by

$$H(\boldsymbol{p}) = \sum_{i=1}^{6} p_i^{\,2}.$$

The value of $H(\boldsymbol{p})$ is obviously maximized at one when every voter has exactly the same preference ranking on candidates with $p_i = 1$ for some $1 \le i \le 6$, and it is minimized at 1/6 for IC with $p_i = \frac{1}{6}$ for all $1 \le i \le 6$. So, this very elementary measure of social homogeneity is totally consistent at its extreme points with what we discussed above. The question of interest is: Does the likelihood that Condorcet's Paradox will be observed really consistently decrease when the level of social homogeneity, as measured by $H(\boldsymbol{p})$, increases?

It was noted above that the probability that Condorcet's Paradox will be observed can be expected simply to be either zero or one, based on $\boldsymbol{p}$, except when we consider the case of DC. Since DC defines the only scenario in which smooth transitions can be observed in the probability that Condorcet's Paradox is observed as $H(\boldsymbol{p})$ changes, we restrict our attention to that case in further discussion. Then, IC remains as the scenario with minimum homogeneity with the restriction of DC, but $H(\boldsymbol{p})$ will be maximized at ½ by Sen's two-class scenario, since the $H(\boldsymbol{p})$ maximizing scenario for the overall case with $p_i = 1$ for some $1 \le i \le 6$ does not meet the requirements for it to be included in DC.

To start our analysis, we need a representation for the probability that Condorcet's Paradox is observed with the assumption of DC in a three-candidate election, which we denote as $P_{CP}^{DC}(n, \boldsymbol{p})$. A formal representation for $P_{CP}^{DC}(n, \boldsymbol{p})$ is obtained in Gehrlein and Lepelley (2011a) for odd n, so that there are no ties in PMR comparisons. The basis of this representation follows from the fact that the probability that any given $\boldsymbol{n}$ will be observed with our urn model for generating random voting situations with independent voter preferences defined by $\boldsymbol{p}$ is obtained directly from the standard multinomial probability $n! \prod_{i=1}^{6} \frac{p_i^{n_i}}{n_i!}$, with:

$$P_{CP}^{DC}(n, \boldsymbol{p}) = 1 - \sum_{m_1=0}^{\frac{n-1}{2}} \sum_{m_2=0}^{\frac{n-1}{2}-m_1} \sum_{m_3=0}^{\frac{n-1}{2}-m_1} \frac{n!}{m_1! m_2! m_3! m_4!} \begin{bmatrix} \left(\frac{1}{2} - p_1\right)^{n-m_2-m_3} p_1^{\,m_2+m_3} + \\ \left(\frac{1}{2} - p_2\right)^{n-m_2-m_3} p_2^{\,m_2+m_3} + \\ \left(\frac{1}{2} - p_3\right)^{n-m_2-m_3} p_3^{\,m_2+m_3} \end{bmatrix},$$

for odd n, where $m_4 = n - m_1 - m_2 - m_3$. No entries for p_4, p_5 and p_6 are required in this probability representation, due to the DC restriction that $p_1 = p_6$, $p_2 = p_5$ and $p_3 = p_4$.

One of the primary difficulties with any analysis of this type has immediately become very evident, since the representation for this very elementary case has already become quite complicated to permit any meaningful analysis, and things will continue to become increasingly more complicated as we progress. This typically leads us to consider only the limiting case for large electorates as $n \to \infty$, and the resulting limiting representation for this probability is obtained as $P_{CP}^{DC}(\infty, \boldsymbol{p})$ in Fishburn and Gehrlein (1980), with

$$P_{CP}^{DC}(\infty, \boldsymbol{p}) = \frac{1}{4} - \frac{1}{2\pi} \sum_{j=1}^{3} Sin^{-1}(1 - 4p_j).$$

Computed values of $P_{CP}^{DC}(\infty, \boldsymbol{p})$ are listed in Table 2.1 and the range of values is truncated since it is obvious that $P_{CP}^{DC}(\infty, \boldsymbol{p})$ does not change if the values of p_1, p_2 and p_3 are permuted. The entries in the Table 2.1 therefore use the smallest value of p_1, p_2 and p_3 for the p_{Min} column index and the mid-range value for the p_{Mid} row index. No table entry is required for the remaining term, since $p_1 + p_2 + p_3 = \frac{1}{2}$ with the assumption of DC.

The limiting probability values in Table 2.1 show that $P_{CP}^{DC}(\infty, \boldsymbol{p})$ goes to zero if any of p_1, p_2 or p_3 is equal to zero, which includes the $H(\boldsymbol{p})$ maximizing two-class scenario for DC. Fishburn and Gehrlein (1980) prove that $P_{CP}^{DC}(\infty, \boldsymbol{p})$ is maximized by the $H(\boldsymbol{p})$ minimizing IC scenario with $p_i = \frac{1}{6}$ for $1 \leq i \leq 6$, and

$$P_{CP}^{IC}(\infty, \boldsymbol{p}) = \frac{1}{4} - \frac{3}{2\pi} Sin^{-1}\left(\frac{1}{3}\right) \approx .08774.$$

An interesting side comment is that a limiting probability representation that directly leads to this equation for $P_{CP}^{IC}(\infty, \boldsymbol{p})$ was first presented in a footnote without any discussion in Guilbaud (1952).

The anticipated relationship between social homogeneity and the likelihood of Condorcet's Paradox holds up at the maximizing and minimizing points of $H(\boldsymbol{p})$, but what happens during the transition between these extreme points? A positively correlated relationship is found to exist between $H(\boldsymbol{p})$ and the probability that

Table 2.1 Computed values of $P_{CP}^{DC}(\infty, \boldsymbol{p})$

	p_{Mid}										
p_{Min}	.000	.025	.050	.075	.100	.125	.150	.175	.200	.225	.250
.000	.000	.000	.000	.000	.000	.000	.000	.000	.000	.000	.000
.025		.041	.048	.051	.053	.054	.055	.055	.056	.056	
.050			.057	.062	.065	.068	.069	.070	.070	.071	
.075				.069	.073	.075	.077	.078	.079		
.100					.077	.080	.082	.083	.083		
.125						.083	.085	.086			
.150							.087	.088			

Table 2.2 Computed values of $E\left[P_{CP}^{DC}(\infty, \boldsymbol{p}|h)\right]$

h	$E\left[P_{CP}^{DC}(\infty, \boldsymbol{p}\|h)\right]$	h	$E\left[P_{CP}^{DC}(\infty, \boldsymbol{p}\|h)\right]$
1/6	.08774	.34	.04394
.18	.08416	.36	.04094
.20	.07787	.38	.03775
.22	.06984	.40	.03431
.24	.05822	.42	.03055
.25	.04746	.44	.02634
.26	.05267	.46	.02141
.28	.05155	.48	.01507
.30	.04932	.50	.00000
.32	.04674		

Condorcet's Paradox is observed in Fishburn and Gehrlein (1980) for the case of all possible $\boldsymbol{p}$ within the assumption of UC being allowed with independent voters, but this relationship tends to deteriorate very significantly as the number of voters gets very large.

When attention is restricted to DC in Table 2.1, it is evident that significant variation still remains in the values of $P_{CP}^{DC}(\infty, \boldsymbol{p})$, so Gehrlein and Roy (2014) took the approach of using an overall view, to look at the limiting expected value of $P_{CP}^{DC}(\infty, \boldsymbol{p})$ with the assumption of DC as $n \rightarrow \infty$, conditional on the fact that $H(\boldsymbol{p})$ is fixed at a specified value h. We denote this conditional expected value as $E\left[P_{CP}^{DC}(\infty, \boldsymbol{p}|h)\right]$ and Table 2.2 lists these expected values for each $h = .18(.02).50$, along with $h = 1/6$. The results of Table 2.2 show a definite overall decline in the expected probability that Condorcet's Paradox will be observed as $H(\boldsymbol{p})$ increases, except for one small region around $h = 1/4$. So, our expectations are strongly supported when attention is restricted to the assumption of DC.

What else do the DC results from Table 2.1 tell us, despite the fact that we do not expect to see anything like DC in actual electorates? It has been concluded that this assumption will produce exaggerated estimates of the probability that paradoxical outcomes that involve PMR relationships will be observed, and IC will give the most exaggerated likelihood estimates of all. Based on Table 2.2, the expected likelihood that Condorcet's Paradox will be observed drops sharply as $H(\boldsymbol{p})$ moves away from the lowest possible level of social homogeneity with IC. Moreover, the significant overestimate of the probability that Condorcet's Paradox will be observed for three-candidates with IC is only approximately 9%. It is quite reasonable to conclude from these inflated values that any such observations should actually be rather infrequent phenomena in real life examples.

Studies that are based on DC and other assumptions like it have never been intended to produce estimates of the probability that Condorcet's Paradox would ever be observed in any actual voting scenario, but they can tell us a great deal about the maximum likelihood of observing paradoxical voting outcomes in extreme cases. Moreover, all of the observations that have been made have only been influenced by making changes in $H(\boldsymbol{p})$, with no additional factors such as dependence among voters' preferences being introduced into the analysis.

2.2.4 Condorcet Efficiency with DC

The Condorcet Efficiency of a voting rule was defined as the conditional probability that the voting rule will elect the CW, given that such a winner exists. And, given that we have now seen very strong evidence that there is a high probability that a CW will exist, it is of obvious interest to consider the probability that voting rules will elect that candidate. Gehrlein (1999) obtains a representation for the limiting Condorcet Efficiency of BR as $n \rightarrow \infty$ with DC, which we denote as $CE_{BR}^{DC}(\infty, \boldsymbol{p})$. This equation starts to become quite complex, so we simply present computed results that are obtained from it in Table 2.3.

For the $H(\boldsymbol{p})$ minimizing case of IC, it is found that $CE_{BR}^{IC}(\infty, \boldsymbol{p}) \approx .90119$ and this is obviously not the scenario that minimizes $CE_{BR}^{DC}(\infty, \boldsymbol{p})$ in Table 2.3. The $H(\boldsymbol{p})$ maximizing case of the two-class society does however maximize the Condorcet Efficiency of BR at a value of one with DC.

As voters' preferences become increasingly homogeneous, intuition suggests that not only should it become increasingly more likely that a CW will exist, but it also seems plausible to think that reasonable voting rules should become increasingly more likely to pick that candidate, so the Condorcet Efficiency of voting rules should tend to increase as voters' preferences become more homogeneous.

This concept has been referred to as the Efficiency Hypothesis in the literature, and we observe some clear violations of this notion for some $\boldsymbol{p}$ with DC in Table 2.3. However, we might be asking too much to expect this relationship hold up precisely for every specific $\boldsymbol{p}$ in DC. It would be more reasonable to consider whether or not the Efficiency Hypothesis is valid on the basis of expected values, just as in the analysis of $E\left[P_{CP}^{DC}(\infty, \boldsymbol{p}|h)\right]$ earlier. Gehrlein and Roy (2014) pursued this by developing a representation for the conditional expected value of the Condorcet Efficiency of BR with DC when $H(\boldsymbol{p})$ has a specified value of h. We denote this as $E\left[CE_{BR}^{DC}(\infty, \boldsymbol{p}|h)\right]$, and Table 2.4 contains computed values that are

Table 2.3 Computed values of $CE_{BR}^{DC}(\infty, \boldsymbol{p})$

	p_{Mid}										
p_{Min}	.000	.025	.050	.075	.100	.125	.150	.175	.200	.225	.250
.000	1.000	.950	.933	.922	.914	.908	.904	.090	.899	.898	.898
.025		.941	.931	.923	.918	.913	.910	.908	.907	.906	
.050			.923	.917	.913	.910	.907	.906	.905	.904	
.075				.913	.909	.907	.905	.904	.903		
.100					.907	.905	.903	.902	.902		
.125						.903	.902	.902			
.150							.901	.901			

Table 2.4 Computed values of $E\left[CE_{BR}^{DC}(\infty,\boldsymbol{p}|h)\right]$

h	$E\left[CE_{BR}^{DC}(\infty,\boldsymbol{p}|h)\right]$	h	$E\left[CE_{BR}^{DC}(\infty,\boldsymbol{p}|h)\right]$
1/6	.90119	.34	.92386
.18	.90229	.36	.92810
.20	.90407	.38	.93279
.22	.90600	.40	.93806
.24	.90784	.42	.94408
.25	.90764	.44	.95113
.26	.91005	.46	.95976
.28	.91320	.48	.97133
.30	.91648	.50	1.00000
.32	.92001		

obtained for each $h = .18(.02).50$, along with $h = 1/6$ and $h = .25$. These results clearly show that the Efficiency Hypothesis is indeed valid for BR on an expected value basis over almost the entire range of $H(\boldsymbol{p})$ values with DC. The one very minor exception is that a minor aberration occurs in the neighborhood of $h = 1/4$, just as we observed earlier in the case with the expected probability that Condorcet's Paradox is observed.

Some evidence has therefore been observed to conclude that increases in social homogeneity, as measured by $H(\boldsymbol{p})$, do generally produce the outcomes on an expected value basis that intuition suggests, by causing both an increased likelihood that a CW exists and an increased level of Condorcet Efficiency of BR. In particular, this has been shown for the limiting case of voters with DC, so these results are driven completely by changes in $H(\boldsymbol{p})$, without any impact whatsoever being caused by the introduction of some dependence among voters' preferences.

2.3 Models with Dependent Voter Preferences

The impetus behind the development of the DC model was to consider cases that maintain an expected perfect balance for individual voter's preferences on pairs of candidates while constructing a voter profile with the urn model that was described for a given $\boldsymbol{p}$. As a result, conditions were not biased to either force a CW to exist or to force the existence of Condorcet's Paradox. We now change our attention to consider a similar type of scenario with regard to maintaining an expected balance among voter's preferences on pairs of candidates for overall voting situations, instead of for each individual voter's preferences.

2.3.1 The Impartial Anonymous Culture Condition

The assumption of IAC is not based on the use of any particular $\boldsymbol{p}$ to generate a random voter profile that is used to obtain a random voting situation. Instead, IAC is based directly on the assumption that each possible voting situation with n voters is equally likely to be observed. IAC also produces an expected perfect balance of preferences on pairs of candidates, just like DC. But, this balance does not apply to the preferences of each of the specific individual voters in a voter profile, it applies over the set of all possible voting situations with anonymous voters.

We used $\Delta(A,B)$ to denote the advantage that a given $\boldsymbol{p}$ gives to Candidate A over B under PMR comparison, and we noted that DC gave a perfect expected balance in comparing A and B since $\Delta(A,B)=0$ for any $\boldsymbol{p}$ meeting the restrictions of the DC assumption. The same basic concept is applied now with respect to a given voting situation $\boldsymbol{n}$, where $D(A,B)$ denotes that number of voters with linear preferences containing $A \succ B$ minus the number of voters with $B \succ A$, and:

$$
\begin{aligned}
D(A,B) &= n_1 + n_2 + n_4 - n_3 - n_5 - n_6 \\
D(A,C) &= n_1 + n_2 + n_3 - n_4 - n_5 - n_6 \\
D(B,C) &= n_1 + n_3 + n_5 - n_2 - n_4 - n_6.
\end{aligned}
$$

We define the *dual voting situation* $\boldsymbol{n}^*$ which reverses the preference ranking on candidates for every voter from the given $\boldsymbol{n}$ with the interchange $n_1 \leftrightarrow n_6$, $n_2 \leftrightarrow n_5$ and $n_3 \leftrightarrow n_4$. When we let $D^*(A,B)$ denote the defined measure of $D(A,B)$ when it is applied to $\boldsymbol{n}^*$, it is obvious from the definitions that $D(A,B)+D^*(A,B)=0$, $D(A,C)+D^*(A,C)=0$ and $D(B,C)+D^*(B,C)=0$. Every possible $\boldsymbol{n}$ can be matched with its unique associated $\boldsymbol{n}^*$, and every one of these voting situations is equally likely to be observed with IAC. It directly follows that the expected values of these terms, when taken over all possible pairs of voting situations, leads to $E[D(A,B)]=E[D(A,C)]=E[D(B,C)]=0$ for any number of voters. So, an expected perfect balance on pairs of candidates holds over the set of all possible voting situations with IAC. If a voting situation is selected at random from the set of all possible voting situations, it is therefore equally likely to have either *AMB* or *BMA* for every possible pair of candidates.

Estimates for the likelihood that voting paradoxes that involve PMR cycles will be observed can therefore be anticipated to be exaggerated with IAC due to this expected balance. However, it is important to stress that the IAC assumption is neither forcing a CW to exist nor forcing Condorcet's Paradox to occur. The DC assumption similarly requires that it is equally likely that *AMB* or *BMA* for all possible pairs of candidates in a voting situation, but this outcome is attained as a result of the much more stringent requirement that it is equally likely to have $A \succ B$ or $B \succ A$ for each individual voter with DC. IAC is more relaxed and does not directly specify anything about the preferences of any individual voter.

The assumption that every voting situation has an equal likelihood of being observed as its dual in some subset of voting situations is not sufficient on its own to conclude that the resulting expected likelihood of observing phenomena like Condorcet's Paradox will be exaggerated for that subset. Consider a scenario in which attention is restricted only to the subset of all voting situations with $n_3 = n_4 = 0$, and add the IAC-like assumption that each of these particular voting situations is equally likely to be observed. The dual of every voting situation in this subset is also included in the subset, so the same arguments that were used above can be used to show that it is equally likely to have *AMB* or *BMA* for all pairs of candidates. But, the rankings in Fig. 2.1 indicate that Candidate *A* is never the middle-ranked candidate for any voter in this subset, so it is either most preferred or least preferred by all voters. If a majority of the odd *n* voters rank Candidate *A* as most-preferred, it must be the CW. The only other option is that a majority of voters rank *A* as least preferred to make it the CL, so that either *B* or *C* must then be the CW. So, a CW must exist for any voting situation in this subset of voting situations with $n_3 = n_4 = 0$, despite the fact that an expected perfect balance is maintained with PMR for all pairs of candidates. Other such conditions can obviously be defined to force the existence of a CW, or to force the existence of Condorcet's Paradox, but neither of these outcomes is being forced to exist if all possible voting situations are equally likely to be observed with IAC.

The IAC assumption is based on a simple concept, and it can take on some other interesting equivalent interpretations. The first of these observations comes from considering UC that is developed in Gehrlein (1981). We have described how the probability that a voting paradox will be observed can be calculated for a specified $\boldsymbol{p}$ that describes the probability that independent voters will have the preference rankings in Fig. 2.1. UC assumes that each such $\boldsymbol{p}$ with $\sum_{i=1}^{6} p_i = 1$ is equally likely to be observed for *n* voters. For any voting paradox, different probabilities will be obtained for observing that paradox with different $\boldsymbol{p}$. But, if we consider the expected value of the probability that this voting paradox will be observed over all possible $\boldsymbol{p}$ with UC, the result will be identical to the probability for *n* voters with IAC. So, IAC has the very nice attribute of giving the expected likelihood that voting paradoxes will be observed. McIntee and Saari (2017) recently consider IAC as a special case of octahedral probability distributions.

Another important interpretation of IAC comes from Berg (1985), who uses *Pólya-Eggenberger* (*P-E*) probability models to show an important connection between IC and IAC. These models are described in the context of generating random voter profiles by drawing colored balls from an urn, following our earlier discussion. The experiment starts with six balls of six different colors being placed into the urn. As before, the color of each ball represents one of the six possible linear preference rankings on candidates. A ball is drawn at random and its corresponding preference ranking is assigned to the first voter. The ball is then replaced, but now *k* additional balls of the same color as the drawn ball are also placed into the urn. A second ball is then drawn, the corresponding ranking for its color is assigned to the second voter, and the ball is replaced along with *k* additional balls of the same color as that second drawn ball. The process is repeated over *n* trials to obtain the preference rankings for the *n* voters. When $k > 0$, the color of the ball that is drawn for the second voter has an increased likelihood of having the same color as the ball that is drawn for the first voter, and so on.

This P-E-based urn experiment describes a general *contagion model* that creates an increasing degree of dependence among the voters' preferences as k increases [See for example Johnson and Kotz (1977)]. When a large value of k is used in these experiments, it becomes very likely that every voter will be assigned the same candidate preference ranking that is assigned to the first voter. However, complete independence exists among voters' preferences for the particular case with $k=0$, which reflects the basis of the general experiment that we used earlier. Since we are starting here with one ball of each color, each preference ranking is equally likely to be selected for the associated voter on each draw, to make this identical to using IC. The particularly interesting observation from Berg (1985) is that the special case that uses $k=1$ in this model is completely equivalent to IAC, so that IAC inherently requires the presence of some degree of dependence among voters' preferences in voting situations. Any comparisons of IC results to IAC results will therefore allow us to consider the impact that introducing a degree of dependence among voters' preferences will have on outcomes, without disrupting the expected perfect balance for PMR comparisons on all pairs of candidates.

2.3.2 *Condorcet's Paradox and Dependent Preferences with IAC*

A representation for the probability $P_{CP}^{IAC}(n)$ that Condorcet's Paradox is observed for an odd number of voters with the assumption of IAC is directly obtainable from a result in Gehrlein and Fishburn (1976), with

$$P_{CP}^{IAC}(n) = \frac{(n-1)(n+7)}{16(n+2)(n+4)}, \quad \textit{for odd } n.$$

Two primary observations can be made from this representation in the limit as $n \to \infty$, where $P_{CP}^{IAC}(\infty) = 1/16$. The first of these observations relates to our earlier discussion of the relationship between the assumptions of IAC and UC. That is, if all possible $\boldsymbol{p}$ vectors are equally likely to be observed, then the expected value of $P_{CP}^{UC}(\boldsymbol{p})$ is 1/16, or about 6%. This again verifies that the IC assumption, which leads to $P_{CP}^{IC}(\infty, \boldsymbol{p})$ equal to about 9%, gives an exaggerated estimate of the probability that Condorcet's Paradox will be observed. Since the proportion of all possible $\boldsymbol{p}$ vectors that precisely meet the restrictions of DC is of measure zero as $n \to \infty$, this observation and previous discussion jointly lead to an alternative form of this conclusion. That is, only 1/16 of all possible $\boldsymbol{p}$ vectors will result in an observation of Condorcet's Paradox as $n \to \infty$, while 15/16 of all possible $\boldsymbol{p}$ vectors will result in the existence of a CW.

The second general observation results from comparing the limiting results that are obtained for the probability that Condorcet's Paradox is observed with IC and with IAC. The limiting probability is approximately 9% with IC, while it is reduced to approximately 6% with IAC. Both of these assumptions were found to result in an expected perfect balance in PMR comparisons on all pairs of candidates in voting situations, so it is natural to wonder what else remains to explain the difference. The earlier discussion of P-E probability models indicated that the difference between IC and IAC stems from the fact that IAC introduces a degree of dependence among voters' preferences, while IC does not do so. As a result, the presence of a degree of dependence among voters' preferences is therefore isolated as being a factor that can reduce the probability that Condorcet's Paradox will be observed.

The degree of dependence that IAC introduces into the preferences of voters decreases as the number of candidates is increased. When there are three candidates, we have $k = 1$ for IAC in the P-E Model, and there is an equal probability of 1/6 for each color to be drawn for the first ball. The probability that the second ball drawn is the same color as the first then increases to 2/7, while the other five colors only have an equal probability of 1/7 of being drawn. The probability that the second ball drawn is a different color than the first is therefore 5/7. The experiment starts with 24 balls of different colors when there are four candidates, and there is a 1/24 chance that each of the colors is drawn to represent one of the possible linear rankings for the first voter's preference ranking. The scenario with $k = 1$ then leads to a probability 2/25 that the second ball drawn is the same color as the first, so there is a probability 23/25 that the second ball drawn is a different color than the first. The probability that there is a difference in colors is significantly greater when there are four candidates, so the degree of dependence correspondingly decreases from the case with three candidates.

The degree of dependence that IAC introduces clearly begins to vanish as the number of candidates becomes large, and any associated IAC probabilities will converge to their corresponding IC probabilities. The rate at which this convergence occurs is illustrated in Table 2.5, with limiting estimates as $n \to \infty$ for the probability that Condorcet's Paradox is observed with IC and IAC for up to six candidates from Feix and Rouet (1999). The differences between the IC and IAC limiting probabilities clearly become very small for as few as six candidates.

2.3.3 *Condorcet Efficiency and Dependent Preferences with IAC*

We saw earlier evidence to give some support for the validity of the Efficiency Hypothesis when the Condorcet Efficiency of BR was generally found to increase as voters' preferences reflected expected increases in their level of social homogeneity, as measured by $H(\boldsymbol{p})$, with the assumption of DC. The presence of statistical

Table 2.5 Limiting probability that Condorcet's Paradox is observed with IC and IAC

		Condorcet's Paradox probability		
Candidates	Rankings	IC	IAC	Difference
3	6	.0877	.0625	.0252
4	24	.1755	.1616	.0139
5	120	.2515	.2477	.0038
6	720	.3152	.3143	.0009

Table 2.6 Limiting Condorcet Efficiency for weighted scoring rules with IC and IAC

λ	$CE_{WSR}^{IC}(\infty,\lambda)$	$CE_{WSR}^{IAC}(\infty,\lambda)$	λ	$CE_{WSR}^{IC}(\infty,\lambda)$	$CE_{WSR}^{IAC}(\infty,\lambda)$
.00	.7572	.8815	.50	.9012	.9111
.05	.7749	.8899	.55	.8984	.8943
.10	.7930	.8979	.60	.8905	.8720
.15	.8113	.9055	.65	.8786	.8461
.20	.8296	.9123	.70	.8639	.8176
.25	.8473	.9182	.75	.8473	.7874
.30	.8639	.9227	.80	.8296	.7560
.35	.8786	.9252	.85	.8113	.7240
.40	.8905	.9249	.90	.7930	.6919
.45	.8984	.9208	.95	.7749	.6603
.50	.9012	.9111	1.00	.7572	.6296

dependence should also be linked to a propensity of voters' preferences to be more alike, so the Efficiency Hypothesis should also apply as the degree of dependence among voters' preferences increases.

The potential relationship between the Efficiency Hypothesis and the existence of dependence among voters' preferences can be tested by considering the differences in Condorcet Efficiency that occur when comparing results that are obtained with IC to those with IAC. Representations for the limiting Condorcet Efficiency of $WSR(\lambda)$ as $n \rightarrow \infty$ with IC were developed in Gehrlein and Fishburn (1978), and the results are denoted as $CE_{WSR}^{IC}(\infty,\lambda)$. It was found that $CE_{WSR}^{IC}(\infty,\lambda)$ is symmetric about $\lambda = 1/2$, with $CE_{WSR}^{IC}(\infty,\lambda) = CE_{WSR}^{IC}(\infty, 1-\lambda)$. BR was found to produce the maximum Condorcet Efficiency, while PR and NPR resulted in having the lowest efficiency. Computed values of these Condorcet Efficiencies are listed in Table 2.6 for each $\lambda = .0(.05)1.0$.

Cervone et al. (2005) later obtained limiting representations for $CE_{WSR}^{IAC}(\infty,\lambda)$, to introduce some dependence among voters' preferences with IAC. The symmetry about $\lambda = 1/2$ was found to disappear with IAC and BR was not found to produce the maximum level of Condorcet Efficiency. Computed values of these efficiencies are listed in Table 2.6, where the results behave exactly as the Efficiency Hypothesis suggests for $0 \leq \lambda \leq .5$, with Condorcet Efficiency values consistently increasing in the transition from IC to IAC. PR is the most dramatic case, where the introduction of some dependence with IAC causes the Condorcet Efficiency to increase from .7572 in the case of complete independence with IC to .8815. The increase for BR only jumps from .9012 to .9111, so the results for BR are much more stable as dependence is introduced. The contrary result occurs for $.5 \leq \lambda \leq 1.0$, where the Condorcet Efficiency decreases in the transition. The efficiency of NPR significantly decreases from .7572 to .6296 when dependence is introduced. While the Efficiency Hypothesis worked quite well on an expected value basis for the highly structures case of DC, it begins to fall apart here. This introduces a phenomenon that we shall see repeated a number of times as we work through this study. Different voting rules may, or may not, abide by the Efficiency Hypothesis, depending on the particular metric that is being used to measure the degree to which voter's preferences tend to be alike.

We have said a lot about how these different probability models can be used to analyze both the likelihood that Condorcet's Paradox might be observed and the Condorcet Efficiency of some common voting rules. The obvious question is: Since we have already admitted that these specific models should never be expected to reflect the conditions of any given election, why do we consider them at all?

2.4 The Relevance of DC, IC, UC and IAC Models

Some recent studies have raised significant concerns about the basic logic of performing any theoretically based research that is related to the likelihood that various voting paradoxes might be observed. This concern was raised after empirically-based analyses of data from actual election results led to the conclusion that the distribution of voters' preferences on candidates in most elections do not correspond to anything like DC, IC, UC or IAC. The most notable study of this type is the work of Regenwetter et al. (2006), and more will be said about the conclusions of this work later. However, further analysis indicates that there are in fact many very good reasons to explain why it is indeed very relevant to consider the results that are obtained with such theoretically based models. A number of arguments that support theoretical investigations that are based on the use of assumptions like DC, IC, UC and IAC are given in Gehrlein and Lepelley (2004):

1. They are very useful when large amounts of relevant empirical data are not available, which is typically the case when analyzing elections.
2. They can show that some paradoxical events are very unlikely to be observed. That is, if we use conditions that exaggerate the likelihood of observing paradoxes to find that the probability for some paradox is small with such calculations, then this paradox is assuredly very unlikely to be observed in reality.
3. They can suggest the relative impact that paradoxical events can have on different types of voting situations. For example, different voting rules can be compared on the basis of their Condorcet Efficiency.

4. By using such probability models to obtain closed form representations, it is easy to observe the impact of varying specific parameters of voting situations or voter profiles, which is more difficult to do with other approaches.
5. The probability representations that are obtained are directly reproducible and verifiable with mathematical analysis, which is not as simple to do with other approaches.
6. Analysis of this type can be useful to find out if the relative probabilities of paradoxical outcomes on various voting mechanisms behave in a consistent fashion over a number of different assumptions about the likelihood that voting situations or voter profiles are observed.

The first point in this list is very critical, since actual elections are almost always conducted with only one voting rule being used, and it typically is not at all easy to compare the resulting election outcome to what else might have happened if some different voting rule had been used. It is not always straightforward simply to determine whether or not the CW has been selected as the winner, based only on the reported vote counts from the election. Assumptions must be made to do this regarding the types of candidate preference rankings that voters might have and to define admissible voting behavior given those preferences. The required severity of these restrictions leads Brams and Fishburn (1983, p. 95) to conclude that due to the wide variety of strategies that voters might use and the lack of full detail about how people voted, the likelihood that it can correctly be concluded that the winner of an election is the CW "is often small if not zero."

As a result of this, other factors about voting behavior must typically be assumed with some model that reconstructs the preferences of voters from the reported ballot outcomes in an election, to establish which candidate was the CW, or to determine what might have happened if a different voting rule had been used. The significant difficulties that can arise from making such assumptions in these models are pointed out in an empirical study by Regenwetter et al. (2002, p. 461), where four actual databases were examined to reconstruct voters' preference rankings with a very basic and plausible model. It was found that very different outcomes could be obtained with only minor changes in a preference threshold parameter that was used in their model. They conclude that "even the most basic and subtle changes in modeling approaches can affect the outcome on any analysis …against the Condorcet criterion."

The obvious conclusion from this observation is that any appeal to dismiss theoretical models as being unrealistic, to instead exclusively pursue empirical studies that are based only on results from typical elections, can lead to an entirely new set of significant issues regarding the validity of the results that are obtained, unless extreme care is taken.

2.4.1 Performing Empirically-Based Evaluations of Theory

Tideman and Plassmann (2014) developed an ingenious solution to the problem that very little information is available regarding voters' preference rankings on candidates in actual three-candidate elections. Their approach assumes that voters' preference rankings on candidates are obtained by a natural process that is based on a *geometric model*. These types of models are routinely used in this type of analysis, and their particular application begins by having each of the three candidates identified by coordinate points in a plane according to the measurable stances that each of them takes on two common attributes of importance to the electorate. Each voter then has their own ideal point in this plane to identify the coordinates that their ideal candidate would take on the two attributes. It is extremely unlikely that the ideal point for any voter would precisely correspond with the exact position points of any of the candidates. So, the voters will then choose their own most preferred candidate as the one that is positioned closest to their ideal point in the plane. The middle-ranked candidate will then be determined as the one that is next to being the closest to their ideal point, and so on.

We demonstrate this geometric approach with the example in Fig. 2.2, where there are two measurable attributes that determine the positions of candidates and voters in a plane. Candidate A has taken a position on the two attributes that corresponds to the coordinates (a_1, a_2), and the other two candidates have done the same with their respective positions. We now consider a randomly selected voter whose ideal candidate would be positioned at coordinates (v_1, v_2), and of course this ideal point does not coincide with any of the three actual candidates. The voter will then select B as the most preferred candidate, since it is the closest candidate in terms of Euclidean distance to the voter's ideal point. Then, C will be the middle-ranked candidate for the voter, and A will be the bottom-ranked since it is the farthest distance from the voter's ideal point.

Earlier studies have used geometric models as a basis to develop computer simulation techniques that generate random voter profiles. This is done by first assuming some type of probability distribution that is used to generate three random position points for the candidates. With the candidate position points remaining fixed, random ideal points are then sequentially generated for each voter according to some specified distribution, and each voter's preference ranking is then constructed with the geometric model from these ideal points to obtain the random voter profile. Tideman and Plassmann (2014) significantly refined this process by starting with data from actual rankings on more than three candidates to calibrate the parameters of the distributions that they use to generate the random candidate positions and ideal points for the case of three candidates. The information that is used to calibrate their model comes from analyzing the initial ranking data to obtain the set of rankings on all triples of candidates from within the original rankings. By doing this, they are able to generate any number of random voter profiles on three candidates that have very similar overall characteristics to the set of triples of candidates from the original rankings. While their procedure ultimately obtains

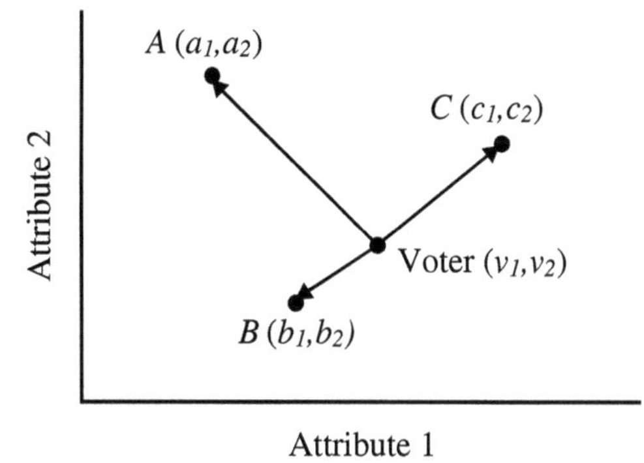

Fig. 2.2 Example to obtain a voter preference ranking with a geometric model

random voter profiles from a computer driven simulation, it is empirically-based since the parameters of the probability models that are used to govern the simulation process are all derived from observed data. We shall be relying extensively on results that are obtained from this *Tideman-Plassmann Model* to provide evidence of support for the conclusions that are drawn from theoretical analysis.

Three primary databases have been used for calibrating the Tideman-Plassmann Model in a number of simulation-based studies. The first is compiled from ranking data from actual elections that were collected by the *Electoral Reform Society* (ERS). The other two were obtained from surveys that used *thermometer scores*. In these surveys, the subjects assigned points to each potential candidate from a range like zero to 100, where larger numbers indicated a greater degree of preference. The candidate ranking for a triple of candidates is then obtained from their associated ranking of thermometer scores from the survey. The second and third databases were both based on 19 surveys each from the *American National Election Studies* (ANES) and from *Politbarometer* (PB) surveys that were conducted by the German Institute for Election Research.

2.4.2 An Empirically Based Evaluation of Preliminary Results

Our theoretical discussion has focused on the probability that Condorcet's Paradox will be observed and on the Condorcet Efficiency of some common voting rules. It is now possible to do a reality check, to see if our observations make any sense, by considering results that are obtained from the Tideman-Plassmann Model with the three databases that are used for calibration. This is done in Gehrlein et al. (2016) for the case of 101 voters to obtain estimates for $P_{CP}^{Base}(101)$ and $CE_{VR}^{Base}(101,\boldsymbol{p})$ for each data Base of ANES, PB and ERS for each Voting Rule of PR, NPR and BR. The results are summarized in Table 2.7

What can we conclude from the results of Table 2.7? The most obvious result is that the probability of observing Condorcet's Paradox is very low for these

Table 2.7 Paradox probabilities and efficiencies from Tideman-Plassmann model

	Base		
Event	ANES	PB	ERS
$P_{CP}^{Base}(101)$	.0137	.0053	.0045
$CE_{PR}^{Base}(101,\boldsymbol{p})$	.8426	.9196	.9353
$CE_{NPR}^{Base}(101,\boldsymbol{p})$	.7417	.8196	.8417
$CE_{BR}^{Base}(101,\boldsymbol{p})$	.9429	.9588	.9613

databases, with estimates of 1.4% or less. And, this was very much to be expected from our finding that the same estimate from IAC, which was known to be significantly exaggerated, had a value of about 6%. This is a prime example of Reasons 2 and 6 above in the list of motivations for why the theoretical models are relevant, and much more extreme examples of this application will be observed later.

A lot of variability exists in the Condorcet Efficiency values of the voting rules across the different databases in Table 2.7, which really is not very surprising at all since the databases are obtained under very different scenarios. So, why would anybody then be surprised at all to find that computed theoretical values for Condorcet Efficiency do not look like the results from any particular actual election? There are in fact some very critical observations that can be made when the Condorcet Efficiencies from the three databases are compared to our theoretical results. The Condorcet Efficiency is always the highest for BR and the lowest for NPR in every case for the databases, which is exactly the same outcome that we observed with IAC, and almost observed with IC where PR and NPR were tied below BR. What really stands out as being important is the observation that the empirically-based results and the theoretical results are all telling us this same story. This is a prime example of Reason 3 above as to why the theoretical models are relevant.

We also see that the Condorcet Efficiency of BR remains very stable over the three databases, compared to the wide variability with PR and NPR. This observation is perfectly consistent with what we concluded regarding the relative Condorcet Efficiency values of these voting rules when changing from IC to IAC in Table 2.6, when we were applying Reason 4 in the list above. Despite the fact that nobody has ever claimed that DC, IC, UC or IAC reflects the reality of any given voting scenario, there is obviously very much that we can learn by studying them.

2.5 The Likelihood of Observing Other Voting Paradoxes

A number of other voting paradoxes were described in Chapter 1, and we now present results that have been obtained to estimate the probability that these paradoxes will be observed with IC and IAC. We shall find in general that the most extreme forms of these paradoxes are expected to be rather rare events in reality.

2.5.1 Borda's Paradox

Two forms of Borda's Paradox were defined in Chapter 1. The Strong Borda Paradox occurs when a voting rule elects the CL. The much more stringent Strict Borda Paradox requires that a voting rule reverses the rankings that are obtained with PMR. It was also stated earlier that BR is completely immune to both forms of this paradox, and it was later proved that BR is distinguished as being the only $WSR(\lambda)$ that has this property for the limiting case for voters as $n \to \infty$. Every other $WSR(\lambda)$ can exhibit both forms of Borda's Paradox, and representations have been obtained for the associated limiting probabilities for each with IC and IAC in a study by Diss and Gehrlein (2012).

Let $P^{IC}_{StBP}(\infty, \lambda)$ denote the conditional limiting probability as $n \to \infty$ that a Strict Borda Paradox is observed with $WSR(\lambda)$ with IC, given that a CW exists. When there are only three candidates, a requirement that a CW exists for an odd number of voters is equivalent to requiring that a strict PMR ranking exists. We define $P^{IAC}_{StBP}(\infty, \lambda)$ in the same fashion, and it is proved that a symmetry exists around the case of BR with $\lambda = 1/2$ for both of these probabilities, since $P^{IC}_{StBP}(\infty, \lambda) = P^{IC}_{StBP}(\infty, 1 - \lambda)$ and the same relationship is also valid with IAC. Computed values of both $P^{IC}_{StBP}(\infty, \lambda)$ and $P^{IAC}_{StBP}(\infty, \lambda)$ are listed in Table 2.8 for each $\lambda = .00(.05).50$.

The results from Table 2.8 indicate that $P^{IC}_{StBP}(\infty, \lambda)$ and $P^{IAC}_{StBP}(\infty, \lambda)$ both decrease as λ increases for the interval $0 \leq \lambda \leq .50$, so that the likelihood of the outcome is maximized by both PR at $\lambda = 0$ and NPR at $\lambda = 1$. However, these probabilities are typically less than 1% in all cases. Given that the IC and IAC scenarios are expected to exaggerate the probability that paradoxical events that involve PMR relationships will be observed, it can easily be concluded that any

Table 2.8 Probabilities for observing a Strict Borda Paradox

λ	$P^{IC}_{StBP}(\infty, \lambda)$	$P^{IAC}_{StBP}(\infty, \lambda)$
.00	.0126	.0111
.05	.0100	.0091
.10	.0077	.0073
.15	.0057	.0056
.20	.0039	.0040
.25	.0024	.0027
.30	.0013	.0016
.35	.0006	.0008
.40	.0002	.0003
.45	.0000	.0000
.50	.0000	.0000

actual observations of a Strict Borda Paradox should be very rare events, which is completely consistent with a survey of empirical studies in Gehrlein and Lepelley (2011b). Since these probabilities are so small, no really significant differences can be observed between the cases of IC and IAC from Table 2.8.

The definition of a Strong Borda Paradox specifies conditions that are not as stringent as the requirements for a Strict Borda Paradox, so it is obvious that it should have a greater probability of being observed. Representations for the limiting conditional probabilities $P_{SgBP}^{IC}(\infty,\lambda)$ and $P_{SgBP}^{IAC}(\infty,\lambda)$ that a Strong Borda Paradox is observed are obtained in Diss and Gehrlein (2012). Similar to our previous observations for a Strict Borda Paradox, $P_{SgBP}^{IC}(\infty,\lambda)$ is symmetric around $\lambda = 1/2$, with $P_{SgBP}^{IC}(\infty,\lambda) = P_{SgBP}^{IC}(\infty,1-\lambda)$, but this symmetry relationship is no longer valid for the probability that a Strong Borda Paradox occurs with IAC. Computed values of $P_{SgBP}^{IC}(\infty,\lambda)$ and $P_{SgBP}^{IAC}(\infty,\lambda)$ are listed in Table 2.9 for all $\lambda = .00(.05)1.00$.

The probabilities in Table 2.9 are obviously greater than the associated probabilities in Table 2.8, and they are maximized with the use of NPR for both IC and IAC, with PR having a marginally smaller probability than NPR for IAC. However, all of these probabilities remain less than 4% in all cases. This indicates that observations of a Strong Borda Paradox should be unlikely events, which is consistent with results from empirical studies that show that they only occur occasionally. The increase in dependence among voters' preferences that is inherent to the IAC assumption reduces the already small probabilities of observing a Strong Borda Paradox with the assumption of IC for all $0 \leq \lambda \leq .5$. But, there are some instances in which the IAC probabilities are marginally greater than the associated IC probabilities when $\lambda > .5$.

Table 2.9 Probabilities for observing a Strong Borda Paradox

λ	$P_{SgBP}^{IC}(\infty,\lambda)$	$P_{SgBP}^{IAC}(\infty,\lambda)$	λ	$P_{SgBP}^{IC}(\infty,\lambda)$	$P_{SgBP}^{IAC}(\infty,\lambda)$
.00	.0371	.0296	.50	.0000	.0000
.05	.0303	.0242	.55	.0001	.0002
.10	.0238	.0192	.60	.0007	.0013
.15	.0179	.0146	.65	.0021	.0033
.20	.0126	.0105	.70	.0046	.0061
.25	.0081	.0070	.75	.0081	.0096
.30	.0046	.0042	.80	.0126	.0136
.35	.0021	.0021	.85	.0179	.0178
.40	.0007	.0007	.90	.0238	.0223
.45	.0001	.0001	.95	.0303	.0269
.50	.0000	.0000	1.00	.0371	.0315

2.5.2 Condorcet's Other Paradox

We defined Condorcet's Other Paradox as occurring if no $WSR(\lambda)$ elects the CW. It is of definite interest to obtain some estimate of the relative likelihood with which this paradox might be observed, since the answer has a highly significant impact on the general wisdom of using a $WSR(\lambda)$ at all. Merlin et al. (2002) obtain a limiting representation as $n \to \infty$ for the conditional probability that a similar event is observed in a three-candidate election, given that a CW exists. They consider the probability that the same candidate that is not the CW will be the winner over the entire range of all possible $WSR(\lambda)$ with $0 \leq \lambda \leq 1$. With the assumption of IC, this limiting probability is estimated to be .01808. Given that IC will tend to create voting situations that have a CW with relatively small PMR margins over other candidates, this yields an estimate for scenarios in which Condorcet's Other Paradox should be much more likely to be observed. And, we find that this probability is still very small under a scenario that is expected to exaggerate it.

Gehrlein and Lepelley (2009) obtain a similar representation for this limiting conditional probability, and find nearly the same numerical result with IC. Moreover, a limiting representation for this probability is also found with IAC, and the resulting conditional probability is reduced to 19/1620 = .01173. So, the already very small IC related probability is further reduced with the introduction of some degree of dependence among voters' preferences with IAC. More relaxed conditions are also introduced to consider probabilities that are more closely associated with the pure definition of Condorcet's Other Paradox as we have defined it, but very little change resulted in the associated probabilities that have just been given. It therefore follows that there should be very little concern that Condorcet's Other Paradox would actually be observed in any three-candidate election.

2.5.3 The No Show Paradox

The example that was presented for the No Show Paradox in Chapter 1 was based on an election that was using NPER, which initially used NPR to rank the three candidates to eliminate the loser, and then carried the two top-ranked candidates to a second phase where the ultimate winner was determined by majority rule. The No Show Paradox can only occur with voting rules that rely on such a two-phase elimination process. The only difference between these elimination procedures is the voting rule that is used to determine the loser that is to be eliminated in the first round of voting.

Probability representations for the limiting probability $P_{NSP}^{IC}(\infty, VR)$ that the No Show Paradox is observed with the assumption of IC are obtained in Lepelley and Merlin (2001) for three voting rules. The analysis includes NPER, along with PER and Borda Elimination Rule (BER). Limiting representations for $P_{NSP}^{IAC}(\infty, VR)$ with

Table 2.10 Probability values for the No Show Paradox

Voting rule	$P_{NSP}^{IC}(\infty, VR)$	$P_{NSP}^{IAC}(\infty, VR)$
PER	.0558	.0408
NPER	.1623	.0425
BER	.0502	.0243

the assumption of IAC are also obtained for both PER and NPER. The final piece of information that we need comes from a representation for $P_{NSP}^{IAC}(\infty, BER)$ that is obtained in Wilson and Pritchard (2007). All numerical results are summarized in Table 2.10.

Occurrences of such Monotonicity Paradoxes in general are very often associated to the presence of a PMR cycle in voting situations. Consequently, it should be expected that the introduction of some degree of homogeneity or dependence in voters' preferences will considerably reduce the vulnerability of these elimination rules to this paradox. This expectation is clearly shown to exist in Table 2.10, where the $P_{NSP}^{IC}(\infty, VR)$ probabilities are significantly greater than their associated $P_{NSP}^{IAC}(\infty, VR)$ probabilities, particularly for NPER. With the exception of the entry for $P_{NSP}^{IC}(\infty, NPER)$, all of these probabilities remain less likely than the probability that Condorcet's Paradox will be observed with both IC and IAC. Conditions that would permit the No Show Paradox to be observed should therefore have a relatively low probability of existing, particularly with PER and BER.

The significant impact that the use of an assumption like IC has on exaggerating these No Show Paradox probability calculations can also be considered from the simple fact that PMR is used on the second stage of all of these elimination rules. The nature of IC will clearly tend to support the generation of voting situations for large electorates such that there will be a relatively close PMR comparison between any two candidates that are carried to this second phase, which enhances the opportunity for either of the two remaining candidates to be selected.

2.5.4 Ostrogorski's Paradox

Probability representations for the likelihood that various forms of Ostrogorski's Paradox are observed are developed in Gehrlein and Merlin (2009a) based on an IC-like assumption that we denote as IC'. That is, each possible assignment of voters' preferences on the m issues, according to Party R or Party L positions, is assumed to be equally likely to be observed. This will tend to result in voting situations in which there is a small relative margin of victory that will determine the Majority Party as $n \to \infty$. Such an expected perfect balance of preferences on issues will then make it significantly easier for paradoxical outcomes to be observed with the majority rule votes on issues, compared to scenarios in which most voters are expected to have the same party membership.

Table 2.11 Probability values for observing Ostrogorski's Paradox

	m		
k	2	3	4
0	.0000	.0104	.0005
1	.5000	.2187	.0594
2	.5000	.5312	.3750
3		.2396	.4406
4			.1245

Representations are obtained for the limiting probability $P_{OP}^{IC'}(\infty, m, k)$ as $n \to \infty$ that the majority rule outcomes on exactly k issues are in agreement with the Majority Party positions in an m-candidate election. It follows that $P_{OP}^{IC'}(\infty, m, 0)$ is the probability that a Strict Ostrogorski Paradox will be observed, which would be a very unsettling outcome. These results then become less and less extreme in the degree to which they are paradoxical as k increases for a given m. Computed values of all possible $P_{OP}^{IC'}(\infty, m, k)$ are listed in Table 2.11 for each value of $m = 2, 3, 4$.

Given the completely balanced nature of IC' in obtaining these probabilities, the maximum agreement values occur for k near $(m+2)/2$. Since we know that these probabilities are exaggerated, it is very clear that the likelihood of observing an extreme Strict Ostrogorski Paradox is very small. Moreover, there is good evidence that stronger versions of a Weak Ostrogorski Paradox are also expected to be relatively rare. Less stringent occurrences of a Weak Ostrogorski Paradox will have greater probabilities of being observed, but such outcomes are not really very paradoxical in nature. It is also found in Gehrlein and Merlin (2009a) that the introduction of any bias to make individual voter's preferences on issues more uniformly consistent with the position of either Party R or Party L will have a significant impact to reduce the probability of observing Ostrogorski's Paradox.

2.5.5 *The Majority Paradox*

Representations for the probability that the Majority Paradox will be observed are obtained in Gehrlein and Merlin (2009b) for the limiting probability $P_{MP}^{IC'}(\infty, m, k)$ as $n \to \infty$ that the majority rule outcomes on exactly k issues are in agreement with the Overall Majority Party positions in an m-candidate election. This study uses the same IC' assumption that was defined in the development of the representations for Ostrogorski's Paradox. Computed values of all possible $P_{MP}^{IC'}(\infty, m, k)$ are listed in Table 2.12 for each $m = 2, 3, 4$.

The computed Majority Paradox probabilities in Table 2.12 have very similar characteristics to what we observed for the Ostrogorski's Paradox probabilities in Table 2.11, so similar conclusions can be drawn. That is, there is strong evidence to suggest that extreme versions of a Majority Paradox can be expected to be quite

Table 2.12 Probability values for observing the Majority Paradox

	m		
k	2	3	4
0	.0000	.0000	.0000
1	.5000	.1623	.0417
2	.5000	.5877	.3750
3		.2500	.4583
4			.1250

rare. While less stringent occurrences of a Majority Paradox will have greater probabilities of being observed, such outcomes are not really very paradoxical.

2.5.6 *The Referendum Paradox*

We follow the general format for describing the Referendum Paradox that was presented in Chapter 1 and consider K voting districts that each have an electorate with n' voters. Each district elects one of the two available candidates by majority rule, and the ultimate winner is the candidate that is chosen as the winner in a majority of the districts. It is necessary to define two variations of IAC-based models in this scenario. In the context of the basic IAC assumption, each voting outcome across all of the total Kn' voters would be assumed to be equally likely. The nature of IAC will generally tend to create a significant degree of dependence among the voters' preferences across voting districts in this scenario. Another option that has been used in the literature is the assumption of IAC^* which only applies the standard IAC to voters' preferences within each of the individual voting districts, and then assumes that there is statistical independence between the preferences of voters who are in different voting districts.

Let $P_{RP}^{IAC^*}(n', K)$ denote the probability that the Referendum Paradox will be observed with the assumption of IAC^*. Feix et al. (2004) obtain values for $P_{RP}^{IAC^*}(\infty, K)$ as $n' \to \infty$ and the results for $K = 3, 5, 7$ are listed in Table 2.13, along with the result for the limiting case $P_{RP}^{IAC^*}(\infty, \infty) = \frac{1}{6}$ as $K \to \infty$ that replicates a result of May (1948). Lepelley et al. (2011) use a Monte Carlo Simulation technique with the assumption of IC to obtain the estimates for $P_{RP}^{IC}(\infty, K)$ in Table 2.13. The introduction of a degree of dependence among voters' preferences within each district with IAC^* clearly reduces the probability that the Referendum Paradox is observed for all K.

In a recent paper, De Mouzon et al. (2016) evaluate the probability of observing the Referendum Paradox under IAC with three districts that have electorates of an equal size. As mentioned above, IAC not only introduces a degree of dependence among voters' preferences within each district as IAC^* does, but it also creates a degree of dependence among voters' preferences across districts. The degree of dependence between the preferences of any two voters is therefore the same,

Table 2.13 Probability values for observing the Referendum Paradox

K	$P_{RP}^{IC}(\infty, K)$	$P_{RP}^{IAC^*}(\infty, K)$
3	.1623	.1250
5	.1814	.1432
7	.1912	.1503
∞	.205	.1667

regardless of whether or not they are in the same district. It is shown that $P_{RP}^{IAC}(n', 3)$ goes to zero as $n' \to \infty$, and it is speculated that the same result is true for any number of districts. It is suggested that the "true" model that reflects the probabilities that have been observed for the Referendum Paradox probably lies in between IAC and IAC^*.

It is worthwhile to note here that IAC atypically does not give an inflated estimate of the probability that the Referendum Paradox will be observed, since $P_{RP}^{IAC}(n', 3)$ goes to zero as $n' \to \infty$. The additional degree of dependence that IAC introduces among voters' preferences across districts beyond the assumption of IAC^* predictably reduces the probability that the Referendum Paradox will be observed. But, the use of IAC actually leads to a scenario in which there is such a high degree of dependence among voters' preferences that the probability of observing the Referendum Paradox vanishes. The selection of the model that is used to represent the probability that voting situations are observed must clearly be evaluated very carefully for each case that is considered. When all of this is coupled with the empirical observations that are mentioned in Chapter 1, these calculated probabilities indicate that the risk that Referendum Paradox might occur cannot be considered to be negligible.

2.6 Adding Additional Internal Structure to Voters' Preferences

We now consider a different election format that merges the idea of ranked voter preferences on candidates with the scenario in which a group wants to accept or reject each of two issues that are being considered. The rejection of an option is denoted with a zero and the acceptance of an option is denoted with a one, so $[0, 1]$ describes the outcome in which the first issue is rejected and the second issue is accepted. There are four possible outcomes that could result from the election: $[0, 0]$, $[0, 1]$, $[1, 0]$ and $[1, 1]$. The format that is used here is not driven by the voters' preferences on individual issues, but by their preferences on the possible pairs of outcomes on the issues, where $[0, 0] \succ [0, 1]$ indicates that a voter prefers to have neither issue accepted to having only the second issue accepted. There are 24 rankings that are possible on these four outcomes, but we now consider the possibility that some logical process exists to explain how voters' preferences are formed, to justify the conclusion that some combinations of these possible rankings could be more likely to be observed together than are other combinations.

To describe one such logical process, suppose that the preference on the pair $[0,0] \succ [0,1]$ is included within preference rankings on outcomes, where the same election result exists for the first issue in both of these outcomes, so the resulting preference relationship is driven by the election result on the second issue. The concept of *Separable Preferences* extends this relationship to hold for all other pairs of outcomes that have the same election result for the first option, to require that $[1,0] \succ [1,1]$ must also be included to maintain consistency in the given rankings. The same general requirement on preferences is also used when comparing pairs of outcomes when the same election result occurs for the second issue, and to cover all similar pairs of outcomes. As a result of this extension to all pairs, while the two initial pairs of outcomes in this example are jointly consistent within the particular ranking $[0,0] \succ [1,0] \succ [1,1] \succ [0,1]$, this ranking must be excluded as a possibility since it is not consistent with the definition of Separable Preferences because it includes $[1,1] \succ [0,1]$, which would require the contradictory result that $[1,0] \succ [0,0]$ within that ranking.

Subsets of eight rankings of these four possible outcomes can simultaneously represent Separable Preferences, and one such subset is given in Fig. 2.3, where r_i denotes the number of voters with the associated preference ranking and $\sum_{i=1}^{8} r_i = n$.

Gehrlein et al. (2017) obtain limiting probability estimates as $n \to \infty$ for the likelihood that Condorcet's Paradox is observed by extending the models of IC and IAC to IC^S and IAC^S that only account for the case when the rankings of the four possible outcomes are restricted to meeting the condition of Separable Preferences with two issues following Fig. 2.3. The results are shown in Table 2.14, along with the associated limiting probabilities that Condorcet's Paradox is observed for IC and IAC with unrestricted preferences.

The results predictably show that the degree of dependence that is introduced with IAC-like assumptions results in a reduced probability of observing Condorcet's Paradox with IAC^S compared to IC^S. It is particularly of interest to note that the introduction of the condition of Separable Preferences also reduces the paradox probability from the unrestricted cases for both IC and IAC. So, scenarios that cause voting situations to reflect models that restrict voter preferences to follow logically consistent patterns of behavior can have the same impact on reducing paradox probabilities as the introduction of dependence among voters' preferences or having increased levels of homogeneity among voters' preferences.

[0,0]	[0,0]	[0,1]	[0,1]	[1,0]	[1,0]	[1,1]	[1,1]
[0,1]	[1,0]	[0,0]	[1,1]	[0,0]	[1,1]	[0,1]	[1,0]
[1,0]	[0,1]	[1,1]	[0,0]	[1,1]	[0,0]	[1,0]	[0,1]
[1,1]	[1,1]	[1,0]	[1,0]	[0,1]	[0,1]	[0,0]	[0,0]
r_1	r_2	r_3	r_4	r_5	r_6	r_7	r_8

Fig. 2.3 Rankings with separable preferences for two-issue outcomes

Table 2.14 Probability for Condorcet's Paradox with separable preferences with two issues

Distribution	Condorcet's Paradox probability
IC	.1755
IC^S	.1667
IAC	.1616
IAC^S	.1250

2.7 Conclusion

A number of voting paradoxes have been considered to obtain estimates of the probability that each might be observed. By using models that are expected to produce exaggerated estimates of these probabilities, we still find that the likelihood of observing any extremely paradoxical outcomes remains very small. At least we find this to be true for the case of three-candidate elections, and we freely admit that increasing the number of candidates will definitely allow for an increased number of avenues through which these paradoxes might occur. But, if a field of many candidates can be reduced with a preliminary process to leave only a small number of the strongest candidates for final consideration, things should then be expected to be quite stable.

The main focus of our analysis has been on the probability that Condorcet's Paradox will be observed. This probability has been found to be expected to be relatively small to begin with, which is strongly supported with evidence from empirically-based simulations. Strong evidence is also provided to indicate that this probability will be further reduced if there is any significant degree of either homogeneity or dependence among voters' preferences. The important point is that since there is a clear expectation that there will be a high probability that a CW exists, the Condorcet Criterion then becomes a very logical criterion to use for evaluating voting rules.

Our theoretical models provide evidence that common voting rules will yield very different levels of Condorcet Efficiency, with BR always showing the best performance of the rules that are considered. Results from empirically-based simulations perfectly reinforce these findings. Moreover, the Condorcet Efficiencies of the different voting rules that are considered are found to change in very different ways as either the level of homogeneity or the degree of dependence among voters' preferences increases.

By considering the impact of requiring Separable Preferences, we have found that there are various restrictions that can be imposed so that voters will form preferences with an increased degree of logical consistency. This additional structure in voters' preferences then leads to a reduction in the probability that Condorcet's Paradox will be observed. Other restrictions of this type will be the primary focus of the next chapter.

References

Berg, S. (1985). Paradox of voting under an urn model: The effect of homogeneity. *Public Choice, 47*, 377–387.

Brams, S. J., & Fishburn, P. C. (1983). *Approval voting*. Boston: Birkhäuser Publishers.

Cervone, D., Gehrlein, W. V., & Zwicker, W. (2005). Which scoring rule maximizes Condorcet efficiency under IAC? *Theory and Decision, 58*, 145–185.

De Mouzon, O., Laurent, T., Le Breton, M., & Lepelley, D. (2016). *The probability of a divided verdict in a U.S. presidential type election*. Toulouse School of Economics Working Paper.

Diss, M., & Gehrlein, W. V. (2012). Borda's Paradox and weighted scoring rules. *Social Choice and Welfare, 38*, 121–136.

Feix, M. R., & Rouet, J. L. (1999). *Un espace des phases électoral et les statistiques quantiques*. Ecole des Mines de Nantes, unpublished manuscript.

Feix, M., Lepelley, D., Merlin, V., & Rouet, J. L. (2004). The probability of conflict in a U.S. presidential type election. *Economic Theory, 23*, 227–258.

Fishburn, P. C., & Gehrlein, W. V. (1980). Social homogeneity and Condorcet's paradox. *Public Choice, 35*, 403–420.

Gehrlein, W. V. (1981). The expected probability of Condorcet's paradox. *Economics Letters, 7*, 33–37.

Gehrlein, W. V. (1999). The Condorcet efficiency of Borda Rule under the Dual Culture Condition. *Social Science Research, 28*, 36–44.

Gehrlein, W. V., & Fishburn, P. C. (1976). Condorcet's paradox and anonymous preference profiles. *Public Choice, 26*, 1–18.

Gehrlein, W. V., & Fishburn, P. C. (1978). Coincidence probabilities for simple majority and positional voting rules. *Social Science Research, 7*, 272–283.

Gehrlein, W. V., & Lepelley, D. (2004). Probability calculations in voting theory: An overview of recent results. In M. Wiberg (Ed.), *Reasoned choices: Essays in honor of Hannu Nurmi* (pp. 140–160). Turku, Finland: The Finnish Political Science Association.

Gehrlein, W. V., & Lepelley, D. (2009, August 18). A note on Condorcet's other paradox. *Economics Bulletin, 29*.

Gehrlein, W. V., & Lepelley, D. (2011a). The Value of Research Based on Simple Assumptions about Voters' Preferences. In Felsenthal & Machover (Eds.), *Electoral Systems: Paradoxes, Assumptions and Procedures* (pp. 171–199). Berlin: Springer Publishing.

Gehrlein, W. V., & Lepelley, D. (2011b). *Voting paradoxes and group coherence: The Condorcet efficiency of voting rules*. Berlin: Springer.

Gehrlein, W. V., & Merlin, V. (2009a). *On the probability of the Ostrogorski Paradox*. www.researchgate.net/publication/257298457_On_the_probability_of_the_Ostrogorski_Paradox.

Gehrlein, W. V., & Merlin, V. (2009b). *The probability of the Majority Paradox*. www.researchgate.net/publication/257298196_The_Probability_of_Observing_the_Majority_Paradox.

Gehrlein, W. V., & Roy, S. (2014). The structure of voters' preferences induced by the Dual Culture Condition. In Fara, Leech, & Salles (Eds.), *Voting power and paradoxes: Essays in honour of Dan Felsenthal and Moshé Machover* (pp. 347–361). Berlin: Springer Publishers.

Gehrlein, W. V., Lepelley, D., & Plassmann, F. (2016). Should voters be required to rank candidates in an election? *Social Choice and Welfare, 46*, 707–747.

Gehrlein, W. V., Le Breton, M., & Lepelley, D. (2017). The likelihood of a Condorcet Winner in the logrolling setting. *Social Choice and Welfare, 49*, 315–327.

Guilbaud, G. T. (1952). Les théories de l'intérêt général et le problème logique de l'agrégation. *Economie Appliquée, 5*, 501–584.

Johnson, N. L., & Kotz, S. (1977). *Urn models and their application*. New York: Wiley.

Lepelley, D., & Merlin, V. (2001). Scoring run-off paradoxes for variable electorates. *Economic Theory, 17*, 53–80.

Lepelley, D., Merlin, V., & Rouet, J. L. (2011). Three ways to compute accurately the referendum paradox probability. *Mathematical Social Sciences, 62*, 28–33.

May, K. (1948). Probability of certain election results. *American Mathematical Monthly, 55*, 203–209.

McIntee, T. J., & Sari, D. G. (2017). Likelihood of voting outcomes with generalized IAC probabilities. *Mathematical Social Sciences, 87*, 1–10.

Merlin, V., Tataru, M., & Valognes, F. (2002). On the likelihood of Condorcet's profiles. *Social Choice and Welfare, 19*, 193–206.

Regenwetter, M., Grofman, B., & Marley, A. A. J. (2002). On the model dependence of majority preference relations reconstructed from ballot or survey data. *Mathematical Social Sciences, 43*, 451–466.

Regenwetter, M., Grofman, B., Marley, A., & Tsetlin, I. (2006). *Behavioral social choice*. Cambridge: Cambridge University Press.

Sen, A. K. (1970). *Collective choice and social welfare*. San Francisco, CA: Holden-Day.

Tideman, T. N., & Plassmann, F. (2014). Developing the aggregate empirical side of computational social choice. *Annals of Mathematics and Artificial Intelligence, 68*, 31–64.

Wilson, M. C., & Pritchard, G. (2007). Probability calculations under the IAC hypothesis. *Mathematical Social Sciences, 54*, 244–256.

Chapter 3
Measures of Agreement in Voters' Preferences

3.1 Introduction

The previous chapter showed that the expected probability of observing Condorcet's Paradox will generally decrease as the degree of homogeneity in voter profiles increases for the limiting case of voters as $n \to \infty$ with DC. But, this overall relationship is weaker for the case of DC when the expected value scenario is abandoned to consider instead all possible individual $\boldsymbol{p}$ vectors in DC. Moreover, DC is based on a knife-edged balance between individual preferences on pairs of candidates that only exists with a probability of measure zero in the space of all possible $\boldsymbol{p}$ that are contained within the assumption of UC.

This analysis proved to be very useful when we were examining the relationship between the probability that Condorcet's Paradox exists and social homogeneity while maintaining independence between voters' preferences. These results are critical, since we really want to apply the Condorcet Criterion to the evaluation of voting rules, and doing this hinges on knowing that scenarios with a high probability of having a CW represent what should be considered to be likely election situations. When we now turn to an analysis of this relationship for more general scenarios than DC, things become much more complicated.

3.2 Homogeneity and Diversity of Preference

Many studies have considered the impact that various measures of the degree to which voters' preferences tend to be homogeneous, or in agreement, will have on the probability that Condorcet's Paradox will be observed. These metrics cover a wide degree of variation in their level of complexity. We describe some of these measures by using the same definitions from the previous chapter regarding the

W.V. Gehrlein, D. Lepelley, *Elections, Voting Rules and Paradoxical Outcomes*,
Studies in Choice and Welfare, DOI 10.1007/978-3-319-64659-6_3

A	A	B	C	B	C
B	C	A	A	C	B
C	B	C	B	A	A
n_1, p_1	n_2, p_2	n_3, p_3	n_4, p_4	n_5, p_5	n_6, p_6

Fig. 3.1 The six possible linear preference rankings on three candidates

distribution of voters' preferences over the six possible linear preference rankings for three candidates $\{A, B, C\}$ as shown in Fig. 3.1.

The definition of $H(\boldsymbol{p})$ in the previous chapter is a *Population Specific Measure of Homogeneity* since it is evaluated on the basis of the p_i characteristics of the population of voters from which specific voter profiles are obtained. For any $\boldsymbol{p}$ with $p_i > 0$ for $1 \leq i \leq 6$, every given voting situation $\boldsymbol{n}$ could be observed with a nonzero probability, so Condorcet's Paradox will therefore be observed with some nonzero probability. This was the motivation to consider the expected probability that Condorcet's Paradox will be observed in the previous chapter.

A more direct approach to this problem is to consider instead *Situation Specific Measures of Homogeneity*, which are based on the n_i terms from given voting situations. Stronger links can be expected between measures of homogeneity and the probability that Condorcet's Paradox is observed by using the n_i terms from voting situations that have been observed than from using the overall p_i terms for a population from which any voting situation could possibly result. The simplest possible Situation Specific Measure of Homogeneity was presented in a study by Felsenthal et al. (1990) that approached this issue from the opposite direction by measuring the degree of *Preference Diversity* in voting situations. To describe this, each preference ranking on candidates in Fig. 3.1 is used to describe the preferences of a coalition of voters, so that every voter in a coalition has the same preference ranking on candidates. The set of coalitions that exist in a voting situation is defined by $K = \{i : n_i > 0\}$. There are for example only two coalitions in a voting situation when $K = \{1, 3\}$ with $n_1 > 0$, $n_3 > 0$ and $n_1 + n_3 = n$. Parameter k denotes the number of coalitions that exist within a voting situation, and the level of diversity of preference that is present in a voting situation increases as more coalitions are represented as k increases.

3.2.1 Diversity of Preference and Condorcet's Paradox

There is an intuitive connection between the value of Parameter k and the probability that a CW exists. If $k \leq 2$ in a voting situation, the rankings in Fig. 3.1 require that some candidate is never ranked as most preferred by any voter, some candidate is never ranked as least preferred by any voter, and some candidate is never middle ranked by any voter. It was proved in the previous chapter that a CW must exist if some candidate is never middle ranked by any voter in a voting situation. So, a CW must exist with certainty if $k \leq 2$. It is very simple to show that a CW must also exist if some candidate is never most preferred or if some candidate is never least preferred in a voting situation. It then follows directly that Condorcet's Paradox can only be observed in voting situations with $k \geq 3$, if each of the three candidates is also ranked in different coalitions as being most preferred, least preferred and middle ranked.

Such a triple of rankings with $K = 3$ has been defined as a *Latin Square*, and there are two possible Latin Squares for the three-candidate rankings in Fig. 3.1 with $K = \{1,4,5\}$ and $K = \{2,3,6\}$. The existence of one of these two Latin Square triples within a voting situation is a necessary condition for Condorcet's Paradox to exist, but it is not a sufficient condition. The probability of observing Condorcet's Paradox must be nonzero for $k = 3, 4$, since it is possible that only one of the two possible Latin Squares could be present in the associated voting situations, while it is not necessarily true that a Latin Square must be present in this case. When $k = 5$, one of the two Latin Squares, but not both, must necessarily be included in any associated voting situation to increase the likelihood of the existence of Condorcet's Paradox. And, both of the possible Latin Square triples must be included when $k = 6$, which creates the maximum number of options for introducing PMR cycles into a voting situation. The probability that Condorcet's Paradox will exist should therefore generally tend to increase as the degree of diversity of voters' preference, as measured by Parameter k, increases.

The strength of this intuitive relationship between Parameter k and the probability that Condorcet's Paradox exists is studied in Gehrlein et al. (2013), where probability representations are obtained by using an extension of IAC that is denoted by $IAC(k)$. This assumes that Parameter k is specified for a given n, and all possible voting situations with that value of k are equally likely to be observed. Representations are obtained for the limiting probability $P_{CP}^{IAC(k)}(\infty)$ that Condorcet's Paradox exists as $n \to \infty$ with $IAC(k)$. Table 3.1 lists the resulting limiting values of $P_{CP}^{IAC(k)}(\infty)$ for each $1 \leq k \leq 6$, and the observed results completely support the intuitive relationship, since $P_{CP}^{IAC(k)}(\infty)$ never decreases as k increases.

3.2.2 *Proximity Measures for Diversity of Preference*

This observed general relationship is encouraging, so we continue to examine this concept on a more refined level. Whenever a large group of voters form their

Table 3.1 Values of $P_{CP}^{IAC(k)}(\infty)$ for $1 \leq k \leq 6$

k	$P_{CP}^{IAC(k)}(\infty)$
1	0
2	0
3	1/40 = .0250
4	1/20 = .0500
5	1/16 = .0625
6	1/16 = .0625

preference rankings, it is likely that there will always be some rankings that represent the preferences of only a small proportion of the population. In some cases it might be quite difficult to understand why these particular rankings might actually represent the preferences of any coalition at all, but they do nonetheless reflect the preferences of those voters. So, how much impact do such outlier voters from small coalitions really have on the possibility that Condorcet's Paradox might be observed?

Suppose that we have a voting situation with five coalitions, so that $k = 5$. Let a represent the proportion of the n voters who are members of the smallest of these five coalitions. If a is very small for large n, the preferences of this small coalition of outlier voters are very likely to be overwhelmed by the preferences of the remaining four coalitions, so that this small coalition is unlikely to have any impact on the election outcome. The probability of observing election outcomes in this particular case really should be quite similar to what is observed in the $k = 4$ scenario. But, as this a starts to increase, the voters in this smallest coalition become less like a small group of outliers, and voting situations that are increasingly more of the nature of a pure five coalition scenario will be observed. The results of Table 3.1 therefore suggest that the probability that Condorcet's Paradox will be observed should increase as a increases for any given k.

A more detailed analysis of this anticipated outcome is performed in Gehrlein et al. (2013) which starts with all voting situations that could have up to six coalitions, and defines α as the minimum proportion of voters who must have their preferences ignored in a voting situation in order to obtain a reduced voting situation with only k coalitions. The bounds on this proportion are $0 < \alpha \leq \frac{6-k}{6}$, and we refer to α as a *Proximity Measure* that defines how close a given voting situation is to having a specified value of Parameter k. If α is small for a voting situation, then only a small proportion of voters must have their preferences ignored in order to obtain a reduced voting situation in which only k coalitions remain. Following the discussion above, the probability that Condorcet's Paradox is observed should generally increase as α increases for any specified value of k.

Representations are obtained for the limiting probability $P_{CP}^{IAC}(\infty, k, \alpha)$ that the minimum proportion of voters that must have their preferences ignored in order to have only k coalitions remaining is given by α, as $n \to \infty$ with the IAC assumption. The values that are obtained from these representations are summarized in the graph in Fig. 3.2.

The results in Fig. 3.2 show that $P_{CP}^{IAC}(\infty, 5, \alpha)$ is constant over the range of feasible α. The remaining probability distributions all start out according to expectation by being minimized at $\alpha = 0$ for each $k = 1, 2, 3, 4$. Then each consistently starts to increase as α gets larger according to our expectations, but we then find a result that is quite contrary to expectation when each of the $P_{CP}^{IAC}(\infty, k, \alpha)$ distributions reaches some maximum value, and then turns around to decrease as α continues to increase over the remainder of its feasible interval.

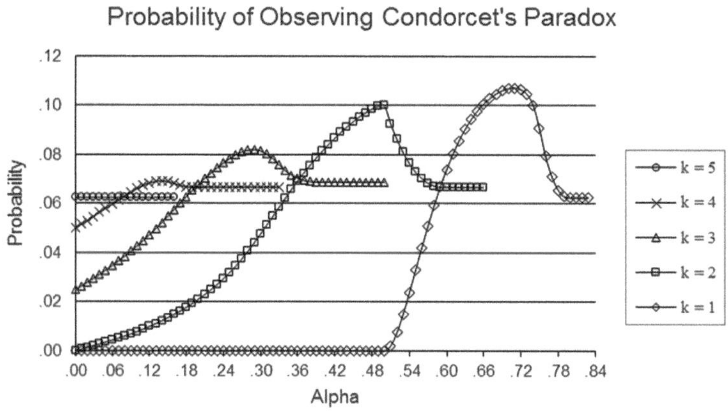

Fig. 3.2 Computed values of $P_{CP}^{IAC}(\infty, k, \alpha)$ for $k = 1, 2, 3, 4, 5$

Parameter k was very appealing as a metric to evaluate the degree of diversity among voters' preferences because of its remarkable simplicity. Things first looked quite promising in Table 3.1 when it proved to be an adequate metric to explain the change in the probability that Condorcet's Paradox exists as the value of k changes in a broad general context. However, Parameter k then disappoints us by failing to act consistently according to expectations when changes in the proximity measure α for a voting situation to have a specified value of k are considered, as observed in Fig. 3.2. We clearly need to develop a more refined measure of diversity or homogeneity.

3.3 Voting Situation Proximity to Meeting Conditions

We want to pursue the idea of finding a simple metric from voting situations that is linked to the probability that Condorcet's Paradox exists that performs much more consistently than what we just observed with Parameter k. The way to do this is linked to earlier discussion when it was suggested that there are some rankings on candidates with small coalitions that might seem difficult to explain, and we pursue this idea further by suggesting that all possible scenarios for a voting situation to have only four coalitions are not equally logical to be observed.

It was mentioned before that a sufficient restriction on three candidate voting situations to require the existence of a CW is that some candidate is never ranked as least preferred by any voter. If this restriction holds, a voting situation has no more than four coalitions because at least one of $n_1 + n_3 = 0$, $n_2 + n_4 = 0$ or $n_5 + n_6 = 0$ holds as a restriction. This is a much more restrictive statement than just saying that there are no more than four coalitions simply because there are at least two of the possible rankings that no voters select to reflect their preferences. What is so special about adding the further restriction that there are no more than four coalitions because some candidate is never ranked as least preferred?

The answer comes from some work in Arrow (1963) where it is shown that this restriction is equivalent to the well-known condition of *single-peaked preferences* from Black (1958). Black developed a natural model that serves as the foundation to explain single-peakedness and proved that the conditions that follow from this model are sufficient to ensure the existence of a CW in three-candidate elections with an odd number of voters. To describe the nature of single-peaked preferences for a general m-candidate election with candidates $\{C_1, C_2, \ldots, C_m\}$, let $U^i(C_j)$ denote the standard measure of *utility* that a voter i associates with Candidate C_j. This utility measure provides a value that a voter associates with having each given candidate win the election, to reflect a relative metric for the voter's degree of preference that is associated with each candidate. Increased measures of $U^i(C_j)$ therefore indicate that a voter has an increased preference for the given candidate, so the given voter's individual preference ranking on candidates will have $C_j \succ C_k$ if, and only if, $U^i(C_j) > U^i(C_k)$.

Consider the simple example voter preference profile for three voters, where each individual voter has a linear preference ranking on six candidates, as shown in Fig. 3.3.

The determination as to whether or not the voter's preference rankings in this profile meet the definition of single-peaked preferences starts by trying to find $U^i(C_j)$ values that are consistent with the preference rankings of the individual voters, while simultaneously meeting an additional restriction. This can be established by drawing a graph like the one that is shown in Fig. 3.4, where values of $U^i(C_j)$ are displayed on the vertical axis of this graph, and the horizontal axis represents a sequence of the candidates that corresponds to a specified overall reference ranking. Let $C_i O C_j$ denote the fact that C_i is ranked before C_j in this overall reference ranking, and the particular overall reference ranking that is used in Fig. 3.4 has $C_2 O C_4 O C_3 O C_6 O C_5 O C_1$.

Figure 3.4 shows plots of these $U^i(C_j)$ values that could possibly be assigned to candidates for each voter to obtain the voter preference rankings from Fig. 3.3. For example, the utility values that are assigned to candidates for Voter 1 have $U^1(C_6) > U^1(C_3) > U^1(C_5) > U^1(C_1) > U^1(C_4) > U^1(C_2)$, to correspond with the linear preference ranking $C_6 \succ C_3 \succ C_5 \succ C_1 \succ C_4 \succ C_2$. These utility values are plotted in the sequence that corresponds to the order of the candidates in the specified overall reference ranking. No claim is being made that these $U^i(C_j)$ values actually represent the precise utility values that voters actually have for the candidates. The only claim being made is that these utilities are possible values that could lead to the observed voters' preference rankings on candidates. Any of the 720 possible linear rankings on the six candidates could have been used as an overall

Fig. 3.3 An example preference profile with three voters and six candidates

Voter 1: $C_6 \succ C_3 \succ C_5 \succ C_1 \succ C_4 \succ C_2$
Voter 2: $C_4 \succ C_3 \succ C_6 \succ C_2 \succ C_5 \succ C_1$
Voter 3: $C_2 \succ C_4 \succ C_3 \succ C_6 \succ C_5 \succ C_1$

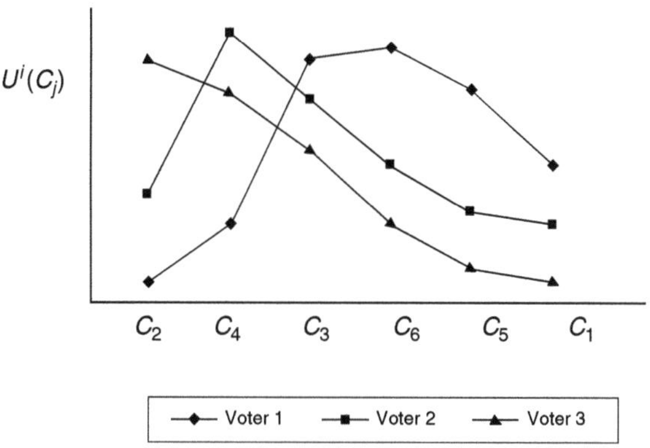

Fig. 3.4 A graph of single-peaked preference curves for three voters

reference ranking, but the specific overall reference ranking that is used for Fig. 3.4 is of particular interest, since it results in plots of the possible $U^i(C_j)$ values that meet Black's definition for single-peaked utility curves for candidates for each voter. Using the original definition in Black (1958, p. 7), a "single-peaked curve is one which changes its direction at most once, from up to down."

This leads to a very natural way to explain how the preferences of the individual voters might have been formed. In particular, this model suggests that all of the voters tend to agree that some underlying characteristics that are associated with the candidates allow for a logical sequencing of the candidates in some natural order from left to right, following the ranking in the overall reference order. Each individual voter then has their own particular most preferred candidate along the spectrum of this sequence of candidates, with decreasing preferences on the candidates as they are ranked farther away, to the left or to the right in the overall reference order, from their most preferred candidate. When we refer to the sequence of candidates from left to right in the overall reference ranking, this does not necessarily indicate a ranking according to the commonly used political terms of 'left' and 'right'. It just refers to a ranking according to the measured values of the characteristics that are being used to evaluate the candidates or alternatives that are under consideration.

A specific example that applies to the scenario described in Fig. 3.4 would consist of a committee of three local government officials who are evaluating six options for building a new school in their district. All members agree that this new school is necessary, but they disagree on how much should be budgeted for building it. Option C_2 represents the least cost "Bare-Bones Scenario" that is adequate for the school to perform the basic required educational objectives with absolutely no additional frills. Option C_1 is the most expensive "Taj-Mahal Scenario" that will not only meet the basic educational requirements, but also provide for many extracurricular activities and advanced educational enhancements. The other options represent budgets for scenarios that fall between these two extremes.

The committee members would obviously agree that the overall reference ranking in Fig. 3.4 would be represented by the increasing costs for the six options from the least cost "Bare-Bones Scenario" to the most expensive "Taj-Mahal

Scenario." Each committee member then has their own personal opinion as to which option, and its corresponding level of spending, best fits the needs of the community. The shape of the utility curves for the individual committee members then reflect the following reasoning. As the budgeted spending levels for options fall farther below any committee member's perceived optimal option, the greater is their concern that the project is falling short of the needs of the community. As the budgeted spending levels for the options increase above their perceived optimal option, the greater is their concern that the project is resulting in excessive spending for a project that exceeds the needs of the community.

Black (1958, pp. 8–9) suggests that this model of single-peaked preferences might be quite plausible to explain voter's preference rankings in many situations. He suggests that such a model is particularly relevant to situations in which committees are comparing options based on "different possible sizes of a numerical quantity and choosing one size in preference to the others." Other natural conditions that are likely to lead to the existence of single-peaked preferences are considered in Buchanan (1970) and Browning (1972). Black then proved that PMR must be completely transitive for an odd number of voters if ***any*** overall reference order and possible $U^i(C_j)$ values that are consistent with the expressed voters' preference rankings can be found to result in single-peaked preference curves for all voters. That is, all voters' preference curves must be single-peaked relative to some overall reference order that is agreeable to all of them.

It is now very clear that not all voting situations that have only four coalitions have the same status. If there are four coalitions only because the electorate views two rankings as being unacceptable, this reduction in preference diversity definitely reduces the probability that Condorcet's Paradox will be observed. However, if there are four coalitions because the voters had preferences that are based on single-peaked preferences, the underlying logic of this model of preference formation adds enough additional coherence to the structure to the preferences in the voting situation to completely eliminate the possibility that Condorcet's Paradox will be observed.

It is not very likely that voting situations will always be perfectly single-peaked with a large electorate, due to the presence of such phenomena as the previously mentioned small coalitions of outlier voters, however small that group might be. Given that background, Niemi (1969) performed a very interesting study to try to explain why Condorcet's Paradox was observed so infrequently in empirical studies with a small number of candidates. The basic premise of this analysis was that voting situations only had to be 'close' to being perfectly single-peaked to have a very high probability that a CW exists. Niemi's measure of 'closeness' of a voting situation to being perfectly single-peaked was simply given as the minimum proportion of all voters whose preferences must be ignored in order for the remaining voters to have perfectly single-peaked preferences. The basis of this analysis was obviously the direct inspiration for the measure α that was used in our earlier analysis that considered proximity measures for preference diversity.

If the preferences of a very small proportion of voters must be ignored in a voting situation to make the reduced voting situation perfectly single-peaked, the likelihood that an occurrence of Condorcet's Paradox will be observed is expected to be very small, since the preferences of that relatively small group can be expected to be overwhelmed by the preferences of the very large proportion of the remaining voters. As this minimum necessary proportion to be removed increases, which means that voting situations are becoming farther removed from having perfectly single-peaked preferences, the probability that Condorcet's Paradox will be observed is then expected to consistently increase.

A formal analysis of this relationship was conducted in Gehrlein (2006), by considering the link between the probability that Condorcet's Paradox occurs and Parameter b that measures a characteristic of voting situations that it is defined by

$$b = \text{Min}\{n_1 + n_3, n_2 + n_4, n_5 + n_6\}.$$

This *Min* function refers to the smallest sum that is contained in the brackets that follow it. With the rankings in Fig. 3.1, Parameter b measures the minimum number of instances in a voting situation that any of the three candidates is ranked as least preferred by voters, and this obviously serves as a perfect surrogate measure for the proximity of a voting situation to being perfectly single-peaked.

We define Parameter b as a measure of the proximity of a voting situation to having a *Perfect Weak Positively Unifying Candidate*. The motivation behind this definition is that when $b = 0$, the electorate is generally unified in support of some candidate in the sense that not a single voter thinks that it would be the worst outcome to have this candidate elected as the winner. A representation that can be used to obtain the limiting probability $P_{CP}^{WPU}(\infty, IAC(\alpha_b))$ that Condorcet's Paradox will be observed when a voting situation has a specified proportion α_b of voters in the electorate who must have their preferences ignored so that a Perfect Weak Positively Unifying Candidate exists is obtained in Gehrlein (2005) with the assumption of $IAC(\alpha_b)$ as $n \to \infty$. These representations are then used to obtain the values for $P_{CP}^{WPU}(\infty, IAC(\alpha_b))$ that are displayed graphically in Fig. 3.5, where the results clearly show that $P_{CP}^{WPU}(\infty, IAC(\alpha_b))$ continuously increases as α_b gets larger over the entire feasible range $0 \leq \alpha_b \leq 1/3$, exactly as intuition suggests.

The use of α to measure the proximity of a voting situation to having a specified Parameter k to reflect preference diversity was not adequate to obtain a consistent pattern of change in the probability that Condorcet's Paradox will be observed. However, the use of Parameter b to measure the proximity of a voting situation to having a Perfect Weak Positively Unifying Candidate, or single-peaked preferences, does indeed perform perfectly according to expectations. The insertion of these additional requirements to suggest that the model for single-peaked preferences could explain why only four coalitions exist in a voting situation has a highly significant impact on our results, relative to the case of saying we only have four coalitions due to a reduction in preference diversity.

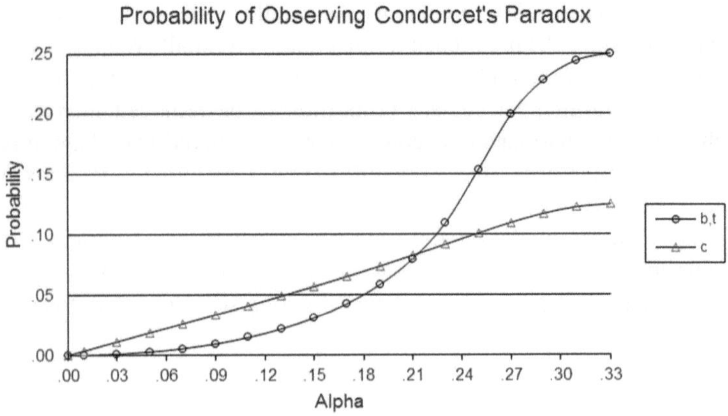

Fig. 3.5 Probability of observing Condorcet's Paradox with weak measures

3.4 Substantive-Level Agreement and Meta-Level Agreement

Very similar arguments have been pursued in the very interesting work of Christian List and his coauthors. The basic concept that is pursued in that body of work starts in List (2002), where it is suggested that different levels of agreement can exist among the preferences of a group of voters. Voters' preferences might indicate a *substantive level agreement* at some elementary level, to the extent that their preferences on candidates tend to reflect some degree of consistency. However, it is further argued that these preferences might go beyond this elementary level of agreement to achieve a higher degree of *meta-level agreement*, to the extent that the voters can additionally agree upon some "common dimension" along which the candidates that are being considered can be conceptualized. It is stressed that while the voters might be largely in agreement as to what this common dimension of comparison is, along with the relative positions of the candidates on that common dimension, they might still be in great disagreement as to what the optimal choice is from the available candidates. This sounds very familiar, since the concept of the voters having mutual agreement concerning the positioning of candidates along such a common dimension is completely consistent with the notion of single-peaked preferences. List (2002) then goes on to argue very effectively that when voters have agreement on such a meta-level, it is significantly more likely that the probability of observing occurrences of paradoxical results like Condorcet's Paradox should be greatly reduced when compared to scenarios in which there is only some elementary agreement on a substantive level.

Dryzek and List (2003) extend this same concept to address the significant complexity that would arise from having multiple relevant issues of consideration that are being used while evaluating candidates, which very well might rule out the possibility of universal agreement among voters on any specific one of the common dimensions that are used for comparing candidates. Recall that the earlier quote from Black provided examples to suggest that single-peakedness would be most appropriate when candidates or alternatives were being evaluated on the basis of a single dimension. However, Dryzek and List go on to suggest that the multiple relevant issue dimensions that are being used to compare candidates could be coupled with the individual voter's preference rankings on candidates on the issue dimensions to lead to some "intra-dimensional single-peakedness." They also discuss the impact that deliberation and discussion might have on voters' "preference structuration" that would increase the likelihood that any resulting voting situation would be more representative of single-peaked preferences.

The impact that learning and deliberation have on levels of agreement among groups of decision makers was evaluated in an empirical study in List et al. (2013), where subjects were formed into groups that were presented with background details from one of 13 different cases. The individuals were polled to obtain their preferences on issues for the case that they were considering both before and after they were given time to have meaningful directed deliberations with other members of their group regarding the relative benefits of issue selection for their case.

The measured level of substantive agreement within each group was actually found to ***decrease*** slightly in the study when voter preferences were compared before and after deliberation, but the proportion of voters in each group that had preferences that were consistent with single-peaked preferences increased as a result of deliberation. This change was strongest for the cases that involved issues that were not perceived as being as critical to the decision makers' personal interests; so that these cases would not have been independently analyzed as thoroughly by the decision makers prior to the deliberations. The increased proximity of the preferences of the decision makers to meeting the condition of single-peakedness that resulted from learning and deliberation therefore tended to be associated with a resulting decrease in substantive level agreement. This all leads to the conclusion that when the subjects deliberated longer, they tended to disagree more on a substantive level. But, greater agreement resulted on a meta-level, since because of deliberation "the more they come to agree about what they are disagreeing about" (List et al. 2013, p. 89).

The conclusion to be reached from these analyses is that there is a definite connection between the degree of voters' understanding of the different 'dimensions' for comparing candidates in an election and their propensity to have single-peaked preferences as a group. The ability of voters to mutually agree on an ordering of candidates along some form of a common dimension leads to a meta-level of agreement that is quite different than simple agreement among preferences on a substantive level. Such a meta-level of agreement among voters then has a significant positive impact on the probability that paradoxical outcomes like Condorcet's Paradox will exist, compared to a scenario with simple substantive level agreement.

A general theory that is related to this topic was presented earlier by Grofman and Uhlaner (1985) that is based on the existence of possible voter "meta-preferences" during the formation of voting situations. It is proposed that a metapreference scenario exists when the voters start out by having mutually agreeable preferences for the characteristics of the broadly defined processes that are to be used in the final determination of their individual preferences on candidates, rather than simply just having preferences on candidates. It is then suggested that voting situations would gain additional structure as a result of this overall understanding of the entire decision process, with such mutually agreeable processes at work in the background while the individual voter preferences are being formed. This increased structure in voting situations would then lead to more overall stability, which would then presumably minimize the likelihood that events such as Condorcet's Paradox would be observed. While suggesting that such a higher level of mutual understanding of the process that leads to the development of voting situations will lead to increased stability in the voting situations, this particular study does not stress the importance of having an underlying scenario in which voters mutually agree upon an ordering of candidates along some form of a common dimension as suggested in the work of List.

The relationship of the very important work of List and his coauthors to the observations that we have made to this point is quite evident. When considering preference diversity with Parameter k, a scenario exists that is strongly suggestive of the presence of an underlying substantive level of agreement among voters, since the resulting voting situations indicate that some of the possible preference rankings are irrelevant, even with $n \to \infty$. This definitely reflects a situation that indicates the existence of some degree of elementary rationality or coherence in the process by which voters are forming preference rankings on candidates, such that voters clearly do not appear to have been simply randomly selecting an ordering on candidates without thinking about it. This substantive level of agreement proves to be adequate to provide logically consistent changes in the probability that Condorcet's Paradox exists as the number of coalitions increases in an overall expected value sense, as seen in Table 3.1. But, preference diversity does not provide an adequate framework to support this same finding in a more refined sense that requires consistent changes in the probability that Condorcet's Paradox is observed as the proximity measure α changes in Fig. 3.2.

Much more is implied when attention is shifted to the consideration of Parameter b, where it is similarly agreed that no more than four of the possible preference rankings are relevant. A much stronger meta-level of agreement is suggested in this scenario, because it is further specified that there are no more than four relevant preference rankings that are feasible in any voting situation ***because*** some candidate does not tend to be ranked as least preferred by any of the voters. Our measure α therefore acts as a proximity measure in a substantive level of agreement scenario, while α_b acts as measure of proximity in a more strongly structured meta-level of agreement scenario. It is therefore quite reasonable to expect results that are related to α_b to behave more consistently than those related to α. Moreover, there is a much stronger relationship between the increases in Condorcet's Paradox probability values (0–.1250) over the range of corresponding α_b in Fig. 3.5 than we observe in the increase in the corresponding probability values (0–.0625) over the range of corresponding α in Table 3.1.

3.5 Measures of Group Mutual Coherence

It is abundantly evident that the use of Parameter b to measure the proximity of voting situations to being perfectly single-peaked, or to have a Perfect Weak Unifying Candidate, works exactly according to expectations regarding the probability that Condorcet's Paradox will be observed. However, the concept of a group decision model that corresponds to single-peaked preferences, reflects just one scenario to describe the behavior of a group that is acting in a mutually coherent, or rational, manner. There are five other such models of *Group Mutual Coherence* that we shall develop, to find that all of them have the same relationship with the probability that Condorcet's Paradox will be observed. They are categorized according being either *Weak Measures* or *Strong Measures*. The proximity of a voting situation to having a Perfect Weak Positively Unifying Candidate has just been examined in detail, and there are two more of these Weak Measures.

3.5.1 Weak Measures of Group Mutual Coherence

A candidate is *Perfect Weak Negatively Unifying* if no voter ranks this candidate as being most preferred, so that the entire electorate is unified in opposition to having this candidate selected as the winner. Parameter t is defined as a measure of the proximity of a voting situation to having such a candidate by counting the minimum number of ballots on which the same candidate is ranked first, with

$$t = \mathrm{Min}\{n_1 + n_2, n_3 + n_5, n_4 + n_6\}.$$

As the value of t decreases in voting situations, there is a smaller proportion of voters that ranks one of the candidates as being the most preferred candidate, and the case with $t = 0$ describes a voting situation in which voters' preferences are perfectly single-troughed or *single-dipped*. Following the logic of the definition of single-peaked preferences in Fig. 3.4, a voting situation meets the condition of having single-dipped preferences if each voter has some least preferred candidate with its minimum utility for all of the candidates along some common reference ordering. Each voter's utility for candidates will then continuously increase for candidates when moving in either direction left or right from their particular least preferred candidate in the common reference ordering.

An example with single-dipped preferences would have voters with utilities on options that look like the example in Fig. 3.4 with all of the utility curves turned upside down. Examples that meet this condition are easy to develop in the context of a "Not in my Backyard Scenario." Consider the case in which it has been determined that a repository for nuclear waste products from hospital and industrial sources is going to be placed near one of six options that represent towns that are located along a highway that runs west to east through a remote region. The facility must be located in the vicinity of one of these towns for access to public utilities, but it will not have a significant positive economic impact on the selected town since it will not employ a large number of people from the area.

Representatives from each of the towns then meet to discuss where this repository should be placed. The overall reference ordering would obviously be the position of the options based on the west to east locations of the associated towns along the highway. It would not be at all surprising to find that each representative would have a least-preferred option that represents the placement of the nuclear waste repository near their own town. Their preferences for options along the west to east positions of the overall reference order would then consistently increase as the options are positioned farther away from their town in either direction.

Following the discussion above for a Perfect Weak Positively Unifying Candidate, the minimum proportion of voters that must be ignored so that the resulting voting situation will have $t = 0$, so that a Perfect Weak Negatively Unifying Candidate exists, is defined as α_t. Gehrlein (2005) develops representations that can be used to obtain the limiting probability as $n \to \infty$ that Condorcet's Paradox will be observed for given values of α_t with the assumption of $IAC(\alpha_t)$. This limiting probability is denoted as $P_{CP}^{WNU}(\infty, IAC(\alpha_t))$, and because of the overall symmetry that exists between the discussion for the logic of the development of single-peaked and single-dipped preferences, it is not at all surprising to learn that $P_{CP}^{WNU}(\infty, IAC(\alpha_t)) = P_{CP}^{WPU}(\infty, IAC(\alpha_b))$ for $\alpha_t = \alpha_b$. All of the discussion about how $P_{CP}^{WPU}(\infty, IAC(\alpha_b))$ changes as α_b increases therefore perfectly applies to the case of considering how $P_{CP}^{WNU}(\infty, IAC(\alpha_t))$ changes as α_t increases.

The use of the two Weak Measures of Group Mutual Coherence that have a foundation in models with voter preferences that are formed on the basis of a common agreement among voters on some underlying reference order on candidates clearly does a great deal to create scenarios in which there is a logical behavior that is observed regarding the probability that Condorcet's Paradox exists. We now consider a model that produces the same kind of logical behavior, without being based on any simple explanation that can be developed based on the existence of some common reference order on candidates.

A *Perfect Polarizing Candidate* exists if no voter has this candidate middle-ranked in their preferences, so that all voters believe that this candidate is either the best or the worst of the three available candidates. Parameter c measures the proximity of a voting situation to describing such a candidate, by representing the smallest number of voters in a voting situation who rank the same candidate in second place, with

$$c = \text{Min}\{n_1 + n_6, n_2 + n_5, n_3 + n_4\}.$$

It was proved in the immediately preceding chapter that a CW must exist for any voting situation if some candidate is never middle-ranked by any voter, which would be the case if $c = 0$. It is obviously not true that every such voting situation must reflect the notion behind the commonly used definition of a truly polarizing candidate, since this happens here when every voter has exactly the same preference ranking; but our particular definition is generally valid in an overall context. We continue to follow the same analysis as before by defining α_c as the minimum proportion of voters that must be ignored in a voting situation so that a Perfect Polarizing Candidate will exist on the reduced voting situation.

Representations from Gehrlein (2005) are used to obtain values for the limiting probability $P_{CP}^{Pol}(\infty, IAC(\alpha_c))$ that Condorcet's Paradox will be observed, given α_c as $n \to \infty$ with the assumption of $IAC(\alpha_c)$. Computed values of this probability are displayed graphically in Fig. 3.5, and it is very clear that the probability results for Parameter c do indeed follow the same logically consistent increase as α_c increases that we observed with Parameters b and t. However, a significantly weaker relationship now exists between the increase in $P_{CP}^{Pol}(\infty, IAC(\alpha_c))$ values (0–.125) over its range of $0 \leq \alpha_c \leq \frac{1}{3}$ than we observed in the increase of $P_{CP}^{WPU}(\infty, IAC(\alpha_b))$ values (0–.250) over the range of corresponding α_b values, and therefore α_t values. The impact of these parameters is clearly different, and it is of interest to analyze what is causing this to occur.

3.5.2 *The Differences Between Parameters b, t and c*

There are some distinct differences that can be observed between the three Weak Measures of Group Mutual Coherence that have been developed. By using Parameter b to measure the proximity of a voting situation to having a Perfect Weak Positively Unifying Candidate, we would expect to observe a reduced level of overall mutual agreement, or *Concordance*, among voters' preferences as b increases. This happens because increasing values of b indicate a movement away from the perfect scenario with $b = 0$ where every voter has a preference ranking in which a given candidate is never viewed as the least preferred. Exactly the same type of thing happens when Parameter t is used to measure the proximity to having a Perfect Weak Negatively Unifying Candidate. That is, as t increases there is movement away from the perfect scenario with $t = 0$ where every voter agrees that some given candidate is never considered to be the most preferred.

On the other hand, our model with a Perfect Polarizing Candidate has a very different nature, since it suggests that voters' preferences will generally reflect changes in levels of *Antagonism,* or divergence of preference, as Parameter c increases away from the perfect case with $c = 0$. Just as we observed with Sen's two-class model, it is not being suggested that holding an election in a highly antagonistic scenario is a desirable condition, but such situations do exist where voters would not be described as acting irrationally while forming their preferences. So, when the phrase Group Mutual Coherence is used, it does not necessarily relate to the degree to which voters' preferences are mutually concordant. It suggests that voters' preferences are being formed on the basis of some underlying process that indicates that the voters are generally acting in a manner that can be exemplified by a model of understandable or rational behavior.

It is of interest to see how levels of concordance and antagonism actually change as Parameters b, t and c change. The classical measure of Concordance from basic statistical analysis is given by Kendall's Coefficient of Concordance (Kendall and Smith 1939), which reduces in our case to $C(\boldsymbol{n})$, with

$$C(\boldsymbol{n}) = \frac{(n_5 + n_6 - n_1 - n_2)^2 + (n_2 + n_4 - n_3 - n_5)^2 + (n_1 + n_3 - n_4 - n_6)^2}{2n^2}.$$

Gehrlein et al. (2016a) performed an analysis to determine how the degree of concordance among voters' preferences, as measured by $C(\boldsymbol{n})$, changes as Parameter b changes, by developing a representation for the limiting conditional expected value $E[C(\infty), IAC|\,\alpha_b]$ of $C(\boldsymbol{n})$ with the assumption of IAC as $n \to \infty$, given that α_b has a specified value. The resulting values of the expected concordance are shown graphically in Fig. 3.6. Similar representations were also obtained for Parameters t and c, and it is found that $E[C(\infty), IAC|\,\alpha_b] = E[C(\infty), IAC|\,\alpha_t]$ when $\alpha_b = \alpha_t$. Those associated computed expected values are also shown in Fig. 3.6, where there is a consistent decrease in the expected value of concordance as each of the Weak Measures α_b, α_t and α_c increases. However, the relationship between the expected concordance for Parameter b, and therefore for Parameter t, is much stronger than the relationship between the expected concordance and Parameter c.

It was noted above that the opposite effect of concordance of voters' preferences is the degree of antagonism that exists among voters' preferences, and Kuga and Nagatani (1974) conducted an analysis of the relationship between the degree of antagonism that is present in voting situations and the probability that an example of Condorcet's Paradox will be observed. Their measure of antagonism is given by $A(\boldsymbol{n})$, with

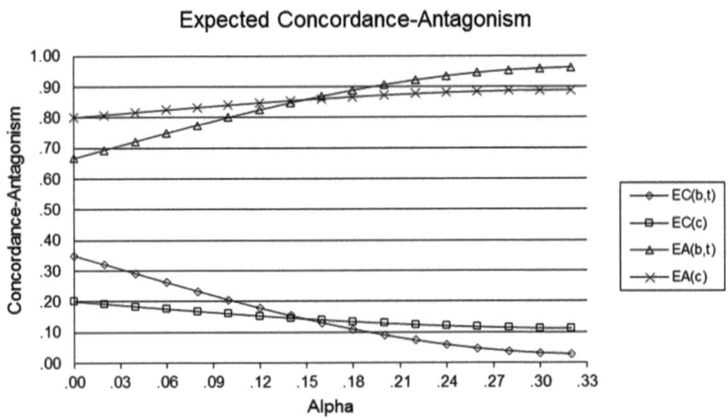

Fig. 3.6 Graphs of expected values of concordance and antagonism

$$A(\boldsymbol{n}) = \frac{4[(n_1+n_2+n_4)(n_3+n_5+n_6)+(n_1+n_2+n_3)(n_4+n_5+n_6)+(n_1+n_3+n_5)(n_2+n_4+n_6)]}{3(n-1)(n+1)},$$

for odd n.

Gehrlein et al. (2016a) followed the logic of the discussion above and developed limiting representations for the conditional expected value of $A(\infty)$ for specified values of Parameters b, t and c. These expected values are denoted by $E[A(\infty), IAC|\,\alpha_b]$, $E[A(\infty), IAC|\,\alpha_t]$ and $E[A(\infty), IAC|\,\alpha_c]$, and it was found that $E[A(\infty), IAC|\,\alpha_b] = E[A(\infty), IAC|\,\alpha_t]$ when $\alpha_b = \alpha_t$. Computed values of the expected antagonism from these representations are shown in Fig. 3.6 for all three Weak Measures.

The expected antagonism consistently increases as each of the Weak Measures increases. As we observed above when considering expected concordance, the relationship that exists between the expected value of antagonism and Parameter b, and therefore Parameter t, is much stronger than the relationship between the expected antagonism and Parameter c. By comparing the probability values for observing Condorcet's Paradox in Fig. 3.5 to the expected values of concordance and antagonism in Fig. 3.6, we find a consistent negative correlation between the likelihood of observing Condorcet's Paradox and the values of expected concordance for all three Weak Measures. This correlation then predictably reverses to be consistently positive between the probability of observing Condorcet's Paradox and the expected level of antagonism for all three Weak Measures.

3.5.3 Weak Measures and Meta-Preferences

The Weak Measures of Group Mutual Coherence that are based on Parameters b and t have been shown to provide sufficient structure to voters' preferences to yield an underlying common ordering of candidates along some dimension, as implied by single-peaked and single-dipped preferences respectively. These two models completely conform to the requirements for meta-level agreement, and they produce the strongest relationship between the associated parameter values and the probability that Condorcet's Paradox will be observed.

We find that Parameter c also provides enough logical structure to voters' preferences to result in a consistent relationship between the parameter values and the probability that Condorcet's Paradox is observed. However, it falls short of requiring the existence of a common ordering of candidates along a common dimension, as suggested for a complete meta-level agreement in the work of List and his coauthors; and we therefore do not observe as strong a relationship between Parameter c and the probability that a Condorcet's Paradox will be observed as we do with the Weak Measures that are based on Parameters b and t.

3.6 Strong Measures of Group Mutual Coherence

Strong Measures of Group Mutual Coherence are based on the number of times that candidates are ranked as most preferred, least preferred or middle-ranked in the expressed preferences of a voting situation, just like the case with Weak Measures. However, Strong Measures provide more direct measurements to describe the inherent structure of voters' preferences. A candidate is for example a *Perfect Strong Positively Unifying Candidate* if every single voter ranks this candidate as most preferred. That indicates a significantly more positive statement about how much the electorate favors this candidate than does the Weak Positively Unifying Candidate that nobody ranks as least preferred. Parameter t^* measures the proximity of a voting situation to having a Perfect Strong Positively Unifying Candidate by determining

$$t^* = \mathrm{Max}\{n_1 + n_2, n_3 + n_5, n_4 + n_6\}.$$

The *Max* function in this definition refers to the largest of the sums that are contained in the brackets that follow it. Things are now reversed from the case of Weak Measures since the ratio $\alpha_{t^*} = t^*/n$ has the range $\frac{1}{3} \leq \alpha_{t^*} \leq 1$, and the closer it is to one the closer the associated voting situation is to having a Perfect Strong Positively Unifying Candidate. That is, the proportion $(1 - \alpha_{t^*})$ of voters must be ignored to have such a candidate in the remaining voting situation. Increasing values of α_{t^*} therefore reflect a stronger overall endorsement of a candidate, and increased levels of concordance, as opposed to the case with a Weak Positively Unifying Candidate where a stronger endorsement was observed in voting situations as values of α_b decreased.

Representations are developed in Gehrlein (2011) that are used to obtain values for the limiting probability $P_{CP}^{SPU}(\infty, IAC(\alpha_{t^*}))$ that Condorcet's Paradox will be observed, for a specified value of α_{t^*} as $n \to \infty$ with the assumption of $IAC(\alpha_{t^*})$. Computed values of these probabilities are displayed graphically in Fig. 3.7 over the feasible range $\frac{1}{3} \leq \alpha_{t^*} \leq 1$.

The probability that Condorcet's Paradox is observed decreases sharply as α_{t^*} increases from one-third, and this probability obviously remains at zero for all $\frac{1}{2} \leq \alpha_{t^*} \leq 1$, since any candidate that is ranked as most preferred by more than half of the electorate in a voting situation must be the CW. There clearly is a very strong connection between Parameter t^* and the probability that Condorcet's Paradox will be observed.

A *Perfect Strong Negatively Unifying Candidate* exists if every voter ranks this candidate as least preferred, which is a much stronger version of a Perfect Weak Negatively Unifying Candidate. Parameter b^* measures the proximity of a voting situation to having such a candidate with

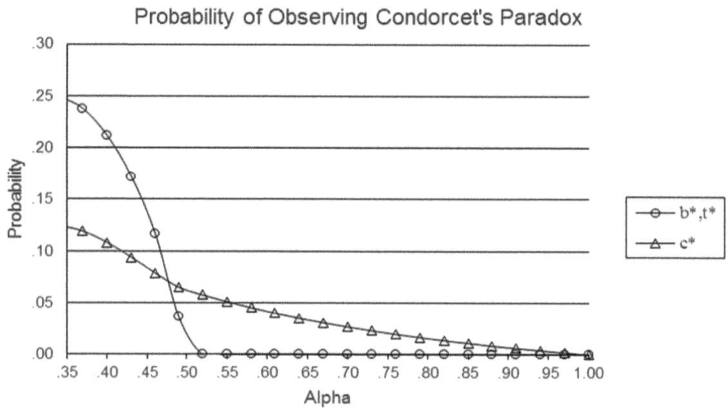

Fig. 3.7 Probability of observing Condorcet's Paradox with strong measures

$$b^* = \text{Max}\{n_1 + n_3, n_2 + n_4, n_5 + n_6\}.$$

Thus, a Strong Measure value of $\alpha_{b^*} = b^*/n$ that is close to one is a much stronger overall rejection of a candidate by the electorate than the rejection that is attributed to a low value for the comparable Weak Measure α_t. Results obtained in Gehrlein (2011) show that $P_{CP}^{SNU}(\infty, IAC(\alpha_{b^*})) = P_{CP}^{SPU}(\infty, IAC(\alpha_{t^*}))$, so the immediately preceding discussion about the relationship between Parameter t^* and the probability that Condorcet's Paradox will be observed also applies to Parameter b^*.

A Perfect Centrist Candidate is middle-ranked by every voter, so that no voter considers this candidate as representing either the best or the worst outcome if elected. Parameter c^* measures the proximity of a voting situation to having such a candidate with

$$c^* = \text{Max}\{n_1 + n_6, n_2 + n_5, n_3 + n_4\}.$$

Following previous discussion, we let $P_{CP}^{Cen}(\infty, IAC(\alpha_{c^*}))$ denote the limiting conditional probability that Condorcet's Paradox will be observed, given a specified value of α_{c^*} as $n \to \infty$ with the assumption of $IAC(\alpha_{c^*})$. Computed values of these probabilities are displayed graphically in Fig. 3.7, where the probability that Condorcet's Paradox is observed consistently decreases as Parameter c^* increases. But, it does not go to zero when a majority of the electorate has the same candidate middle-ranked in their preferences, as it happens with Parameters b^* and t^*. Just as we observed with Weak Measures, the proximity measure that is based on the number of middle-ranked candidates does not have as strong an impact on the probability of observing Condorcet's Paradox as the proximity measures that are based on the count of most-preferred or least-preferred positions in voters' rankings on candidates in a voting situation.

3.7 Refined Measures of Group Mutual Coherence

The motivation behind the consideration of refining the parameters that are associated with the six Measures of Group Mutual Coherence that have been developed so far stems from example voting situations like the one with 37 voters that is shown in Fig. 3.8.

Suppose that you have not been shown this actual voting situation and that you only know that there are 37 voters. Then, somebody tells you that Parameter b for this voting situation has a value of $\text{Min}\{9+5, 1+10, 1+11\} = 11$ and asks your opinion of the relative likelihood that Condorcet's Paradox exists for the example. Since this result only tells you that a minimum of 11 of the 37 voters must be ignored to have a Perfect Weak Positively Unifying Candidate, the odds tell you that the likelihood of observing Condorcet's Paradox is high on a relative basis. Suppose that you are told instead that Parameter t has a value of $\text{Min}\{9+1, 5+1, 10+11\} = 6$. You would now respond that since only six of the 37 voters must be ignored to have a Perfect Weak Negatively Unifying Candidate, the odds that Condorcet's Paradox will be observed have been reduced, but that likelihood is still not near the low end of the spectrum. Finally suppose that you are told that Parameter c has a value of $\text{Min}\{9+11, 1+1, 5+10\} = 2$, so that only two of the 37 voters must be ignored to have a Perfect Polarizing Candidate. Your assessment would then be that the probability of observing Condorcet's Paradox has become relatively small, but not certain, on the basis of that additional piece of information. We get three different answers with the three given parameters values from the same voting situation!

The moral of this story is that while we have seen a consistent relationship between the measures of Group Mutual Coherence and the probability that Condorcet's Paradox is observed, it is very important to know which of these parameters might be serving as the primary driving force to possibly allow an occurrence of Condorcet's Paradox in any particular voting situation. In our example, the fact that Parameter c is so small would clearly override the importance of the observed values of Parameters t and c.

This issue is addressed in Gehrlein and Lepelley (2016) where the analysis begins by looking for *Refined Parameters* that are associated with the Weak Measures b, t and c that directly account for the fact that it makes a real difference as to which of these parameters is the minimum of the three. As an indication of the degree to which this distinction makes a difference, the limiting probability that Condorcet's Paradox is observed with IAC is found to be $\frac{5}{128} = .0391$ when attention is restricted only to voting situations in which either b or t is the minimum

A	A	B	C	B	C
B	C	A	A	C	B
C	B	C	B	A	A
$n_1 = 9$	$n_2 = 1$	$n_3 = 5$	$n_4 = 10$	$n_5 = 1$	$n_6 = 11$

Fig. 3.8 An example voting situation

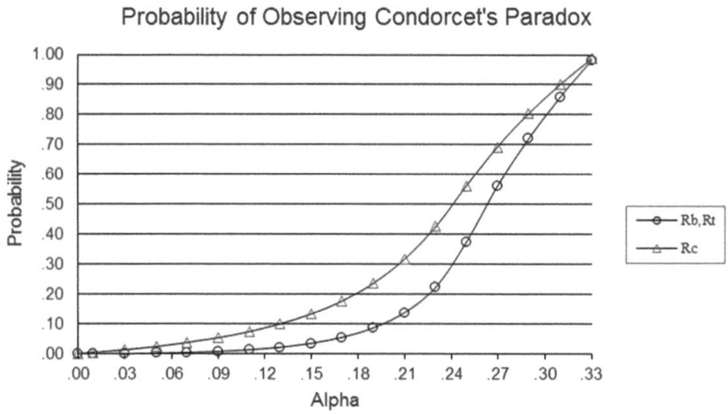

Fig. 3.9 Probability of observing Condorcet's Paradox with refined weak measures

of the three parameters, while it increases dramatically to $\frac{7}{64} = .1094$ when c is the minimum.

A refined version of Parameter b is denoted as Parameter Rb, which takes on the same value as Parameter b for all voting situations in which b is the primary driving force as the $\text{Min}\{b,t,c\}$, and it is undefined for any voting situations in which either of Parameters t or c is the $\text{Min}\{b,t,c\}$. Parameters Rt and Rc are defined in the same fashion and limiting representations are developed that can be used to obtain the conditional probability that Condorcet's Paradox will be observed with IAC as $n \to \infty$, given that the proportions α_{Rb}, α_{Rt} and α_{Rc} are specified. The resulting computed values for each of $P_{CP}^{WPU}(\infty, IAC(\alpha_{Rb}))$, $P_{CP}^{WNU}(\infty, IAC(\alpha_{Rt}))$ and $P_{CP}^{Pol}(\infty, IAC(\alpha_{Rc}))$ are displayed graphically in Fig. 3.9 where it is seen that $P_{CP}^{WPU}(\infty, IAC(\alpha_{Rb})) = P_{CP}^{WNU}(\infty, IAC(\alpha_{Rt}))$.

It is not surprising to see that the probability that Condorcet's Paradox is observed consistently increases as the refined Parameters Rb, Rt and Rc increase over the range of possible parameter values in Fig. 3.9. The surprising factor is that the strength of this relationship is dramatically magnified by using the Refined Weak Measures. Note for example that the range of values for $P_{CP}^{WPU}(\infty, IAC(\alpha_b))$ is 0–.250 in Fig. 3.5 and that this range significantly expands for $P_{CP}^{WPU}(\infty, IAC(\alpha_{Rb}))$ in Fig. 3.9 to cover the entire spectrum of probability values from 0 to 1. The range of probabilities for observing Condorcet's Paradox was also different in Fig. 3.5 for Parameter c, compared to Parameters b and t, since its range was only 0–.125, which expands to the entire possible range 0–1 for the refined Parameter Rc. So, the Refined Weak Measures all generally behave in a more similar manner than the original Weak Measures, since they all have the same range of probability values. But, they still do not all behave in precisely the same manner.

Following the discussion of the refinement of the Weak Measures, Refined Strong Measures for b^*, t^*, and c^* were developed in Gehrlein and Lepelley (2016) that directly account for the fact that it makes a difference as to which of these parameters is the maximum. The significance of this distinction is again made very evident by the fact that when attention is restricted only to voting situation in which either b^* or t^* is the maximum of the three parameters, the limiting probability that Condorcet's Paradox is observed with IAC is only $\frac{7}{256} = .0273$, while it increases by a factor of nearly 5 to $\frac{17}{128} = .1328$ when c^* is the maximum.

The refined parameters for the Strong Measures are then defined for example by Rb^* that has the same value as b^*, conditional on the fact that b^* is the $\text{Max}\{b^*, t^*, c^*\}$. Representations were obtained to allow the calculation of the limiting probability that Condorcet's Paradox will be observed as $n \to \infty$ with IAC for specified values of Parameters Rb^*, Rt^* and Rc^*, and computed values from these $P_{CP}^{SPU}(\infty, IAC(\alpha_{Rb^*}))$, $P_{CP}^{SNU}(\infty, IAC(\alpha_{Rt^*}))$ and $P_{CP}^{Cen}(\infty, IAC(\alpha_{Rc^*}))$ representations are displayed graphically in Fig. 3.10 where we find $P_{CP}^{SPU}(\infty, IAC(\alpha_{Rt^*})) = P_{CP}^{SNU}(\infty, IAC(\alpha_{Rb^*}))$

Just as we observed with the Refined Weak Measure, the relationship between parameter values and the probability that Condorcet's Paradox will be observed is significantly magnified with the Refined Strong Measures. The overall range of these probability values also becomes identical for all three Refined Strong Measures, while this was not true of the original definition of the Strong Measures.

These refined definitions for both Weak and Strong Measures result in a much stronger relationship between the probability that Condorcet's Paradox is observed and the parameter values. The patterns of behavior also become more consistent between the parameters that are included in both the Weak and Strong Measures when this refinement is implemented, but some differences still remain.

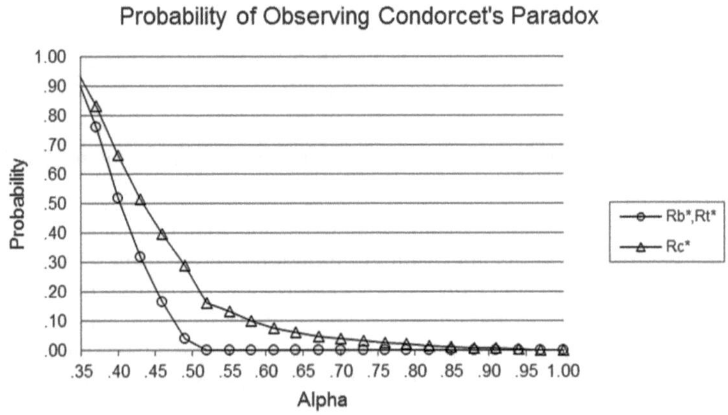

Fig. 3.10 Probability of observing Condorcet's Paradox with refined strong measures

3.8 Conclusion

All of this indicates that it really matters which of the Weak Measures and Strong Measures is the dominating factor in determining the likelihood that a CW exists, so we should focus on looking at the Refined Measures of Group Mutual Coherence in order to keep attention on the primary force that is driving the results. Moreover, each of the models that lead to these Refined Measures could make sense in different scenarios. So, which measures tend to reflect what is most typically observed in actual elections?

An answer to this question is given in Gehrlein et al. (2016b) where a summary is presented from the three databases that were used as the empirical basis to load parameters for the Tideman-Plassmann Model for simulating voting situations that was described in Chap. 2. The relevant results are shown for Weak Measures in Table 3.2, which lists the percentage of times that Parameter $x \in \{b, t, c\}$ is the Min $\{b, t, c\}$ for all triples of candidates from the rankings in the original ANES, PB and ERS databases. Table 3.3 shows similar values for Strong Measures, when each Parameter $x^* \in \{b^*, t^*, c^*\}$ is the Max$\{b^*, t^*, c^*\}$. These are not simulated percentages, they are the actual percentages from the raw databases. The symmetry of IAC with respect to candidates makes all of these values equally likely for that case, ignoring the possibility of ties for a large electorate.

There are some very clear indications that there are distinct differences between these databases regarding the distribution of both the Refined Weak and Refined Strong Measures. The results from the ANES database are much closer to IAC than either of the PB or ERS results. Both PB and ERS are indicative of scenarios in which there is a much greater level of concordance among voters' preferences, with a definitely preferred candidate having a large value of t^* and/or a heavily disliked

Table 3.2 Distribution of parameters for weak measures from the empirical databases

	Base			
Min$\{b, t, c\}$	IAC (%)	ANES (%)	PB (%)	ERS (%)
b	33.33	34.14	42.55	39.30
t	33.33	37.94	43.84	38.74
c	33.33	26.72	11.45	10.92
Tie		1.21	2.17	11.02

Table 3.3 Distribution of parameters for strong measures from the empirical databases

	Base			
Max$\{b^*, t^*, c^*\}$	IAC (%)	ANES (%)	PB (%)	ERS (%)
b^*	33.33	39.15	38.40	40.96
t^*	33.33	29.31	48.93	45.63
c^*	33.33	30.89	11.68	7.81
Tie		0.65	0.99	5.59

candidate having a large value of b^*. The percentage of times that a voting situation most closely represents a scenario either with a highly Polarizing Candidate that has a value of Parameter c that is the minimum of the Weak Measures, or with a strong Centrist Candidate that has a value of c^* that is the maximum of the Strong Measures typically represent the least likely case among their respective groups of parameters. But, these percentages are still sufficiently large that the possibility of observing results for which these are the dominating models cannot be ignored.

Studies of actual databases from the literature indicate that there is only a very small probability that Condorcet's Paradox will ever be observed in practice with a small number of candidates, but that it does happen on occasion. Our analysis shows that this outcome is exactly what should be expected for the case of three candidates, as long as a reasonably large percentage of the electorate is commonly forming preferences that meet the conditions of any one of the six plausible and very simple models of Group Mutual Coherence. This is not good news for anyone who might want to dwell on the small possibility that such a disruptive outcome might possibly be observed. But, it is very good news for those who like the idea of evaluating voting rules on the basis of their propensity to elect the CW, since it is very likely that such a candidate will exist for a small number of candidates.

References

Arrow, K. J. (1963). *Social choice and individual values* (2nd ed.). New Haven CT: Yale University Press.

Black, D. (1958). *The theory of committees and elections*. Cambridge: Cambridge University Press.

Browning, E. K. (1972). A note on cyclical majorities. *Public Choice, 12*, 111.

Buchanan, J. M. (1970). *The public finances*. Homewood, IL: Richard D Irwin Press.

Dryzek, J. S., & List, C. (2003). Social choice theory and deliberative democracy: A reconciliation. *British Journal of Political Science, 33*, 1–28.

Felsenthal, D. S., Maoz, Z., & Rapoport, A. (1990). The Condorcet-efficiency of sophisticated voting under the plurality and approval procedures. *Behavioral Science, 35*, 24–36.

Gehrlein, W. V. (2005). Probabilities of election outcomes with two parameters: The relative impact of unifying and polarizing candidates. *Review of Economic Design, 9*, 317–336.

Gehrlein, W. V. (2006). *Condorcet's paradox*. Berlin: Springer Publishing.

Gehrlein, W. V. (2011). Strong measures of group coherence and the probability that a pairwise majority rule exists. *Quality and Quantity, 45*, 365–374.

Gehrlein, W. V., & Lepelley, D. (2016). Refining measures of group mutual coherence. *Quality and Quantity, 50*, 1845–1870.

Gehrlein, W. V., Lepelley, D., & Moyouwou, I. (2013). The impact of voters' preference diversity on the probability of some electoral outcomes. *Mathematical Social Sciences, 66*, 352–365.

Gehrlein, W. V., Lepelley, D., & Plassmann, F. (2016a). Should voters be required to rank candidates in an election? *Social Choice and Welfare, 46*, 707–747.

Gehrlein, W. V., Lepelley, D., & Plassmann, F. (2016b). *To rank or not to rank: A summary*. Note at: www.researchgate.net/publication/306276950_To_Rank_or_Not_to_Rank_A_Summary.

Grofman, B., & Uhlaner, C. (1985). Metapreferences and reasons for stability in social choice: Thoughts on broadening and clarifying the debate. *Theory and Decision, 19*, 31–50.

Kendall, M. G., & Smith, B. B. (1939). The problem of *m* rankings. *Annals of Mathematical Statistics, 10*, 275–287.

Kuga, K., & Nagatani, H. (1974). Voter antagonism and the paradox of voting. *Econometrica, 42*, 1045–1067.

List, C. (2002). Two concepts of agreement. *The Good Society, 11*, 72–79.

List, C., Luskin, R. C., Fishkin, J. S., & McLean, I. (2013). Deliberation, single-peakedness, and the possibility of meaningful democracy: Evidence from deliberative polls. *The Journal of Politics, 75*, 80–95.

Niemi, R. G. (1969). Majority decision-making under partial unidimensionality. *American Political Science Review, 63*, 488–497.

Chapter 4
Single-Stage Election Procedures

4.1 Introduction

It would be difficult to argue that some candidate other than the CW should be chosen as the winner of an election, when a majority of voters prefer the CW to every other candidate. So, the Condorcet Criterion certainly sounds like a very valid basis for comparing election procedures, as long as a CW exists. We have also found that there are a number of logical models of Group Mutual Coherence that will result in a high probability that a CW will exist with a small number of candidates, as long as a reasonable proportion of the electorate is mutually forming preference rankings on the candidates in accordance with one of these models of rational behavior. The obvious conclusion is that there is very good reason to use the Condorcet Criterion as a basis for evaluating voting rules.

The introduction in the first chapter made it quite evident that over the course of many years, a lot of people have had much to say about what kind of voting rule makes the most sense for use. Most of the voting rules that are of particular interest for three-candidate elections have already been developed, with PR and NPR being the simplest rules of all. Then, the $WSR(\lambda)$ were considered, with particular emphasis on BR. The elimination rules PER and NPER were also defined.

The evaluation of these different rules starts with a consideration of what each of them requires as input from voters, how transparent and understandable the process is for the electorate and the resources that are required to implement each rule. The primary distinction that exists between the voting rule definitions is whether they are single-stage rules or two-stage rules. Single-stage rules elect a winner directly from the information that is provided on the initial ballots that the voters submit, including PR, NPR and BR. The two-stage rules are comprised of the elimination procedures that require voters to participate in two separate elections, and that will be the topic of the next chapter.

W.V. Gehrlein, D. Lepelley, *Elections, Voting Rules and Paradoxical Outcomes*, Studies in Choice and Welfare, DOI 10.1007/978-3-319-64659-6_4

4.2 To Rank or Not to Rank

In terms of the required input from voters to implement single-stage voting rules, one might think that there is no major distinction between the demands of PR, NPR and BR. But, it turns out that very significant opposition has been raised against the simple concept of requiring voters to rank their preferences on candidates, which would eliminate BR from consideration. Some of this opposition has already been hinted at in Condorcet's work in Chap. 1, where he strongly dismissed the credibility of BR as a voting rule. Condorcet further wrote at length regarding his opposition to using any $WSR(\lambda)$ in the example in Fig. 1.5, which led to Condorcet's Other Paradox. It is emphasized that (Condorcet 1788a, p. 151) the failure of a $WSR(\lambda)$ to select the CW in this example was not the fault of the fact that BR was the particular case that was being used, but that "whatever revisions we introduce, this method will always give the wrong result."

Condorcet's general thoughts about the necessity of developing election procedures that will minimize the possibility of making mistakes in election outcomes initially casts a significant shadow on the notion of obtaining preference rankings from voters. In particular, he states (Condorcet 1788b, p. 159) that voting procedures must be established to have voters "expressing their opinions only with a yes or a no." However, Condorcet did suggest that rankings that are obtained from voters could be used to directly determine the CW from the resulting PMR comparisons (Condorcet 1788a, p. 152). He went so far as to develop a proposed voting rule that required voters to submit a ranking on 20 candidates (Condorcet 1788c, p. 141). So, Condorcet's objections to obtaining rankings from voters is specifically focused on his concerns about continuing any further to use any $WSR(\lambda)$ to determine a winner from those rankings.

Dodgson was mentioned earlier as an important nineteenth century researcher on the subject of voting procedures, and he also wrote about the desirability of keeping voting rules as simple as possible. He was involved in an ongoing debate with The Proportional Representation Society regarding a voting rule that the Society had proposed. Dodgson clearly stated his very strong opposition to the requirement of having voters report rankings on candidates in order to implement the proposed procedure (Dodgson 1884, pp. 29–30):

> One great objection to this (proposed) method is the confusion it would cause in the mind of an ignorant Elector, who, though quite able to name his favourite Candidate, would be utterly puzzled if told to arrange 5 or 6 names in order of merit.

Another comment from Black (1958, p. 182) provides some additional input regarding the desirability of requiring preference rankings from voters, when he concludes that the basic notion of obtaining ratings for the relative degrees of preference of decision makers in measurable scales like 3:1 or 4:1 is plausible for the consideration of things like goods in markets. However, he continues on to reveal that he does not believe that the human mind operates in such a way as to allow for the same type of relative evaluation of candidates in an election. This statement definitely falls short of being a complete condemnation of requiring simple ordinal rankings from voters, but it strongly indicates that requiring anything from voters beyond that, such as indicating their relative intensity of preference in comparisons of candidates, would not be appropriate.

4.3 Increases in Efficiency from Using Ranked Preferences

We evaluate the potential benefits that can be obtained from requiring voters to rank candidates so that BR can be used, to determine if it will result in a significantly increased Condorcet Efficiency relative to PR and NPR. This is done despite the concerns that are expressed above by Condorcet regarding the use of any $WSR(\lambda)$ in particular and by Dodgson regarding the use of any rankings in general, since the impact of their arguments is somewhat minimized by the fact that we are limiting attention to elections with only three candidates.

Some preliminary insight to this analysis was given in Table 2.6, which listed the limiting Condorcet Efficiency values as $n \rightarrow \infty$ for $WSR(\lambda)$ with IC and IAC. Some of those values are replicated for convenience in Table 4.1, for the particular cases of PR, NPR and BR.

The pattern of change in Condorcet Efficiency is clearly evident in Table 4.1 when a degree of dependence is introduced among voters' preferences with IAC compared to IC. The efficiency of BR remains quite stable, while PR shows a significant increase and NPR shows a significant decrease. These changes in overall efficiency indicate that the introduction of an increased degree of dependence among voters' preferences starts to make PR look quite favorable compared to BR, while making NPR appear not to be a viable option. This actually brings into question the value of any expected benefit that might be obtained from requiring voters to provide ranked preferences. Why should we act against the concerns that have been expressed about requiring voters to provide rankings, when PR is showing such significantly improved performance relative to BR when some dependence is introduced?

This phenomenon is further examined in Gehrlein (2003) in the context of the P-E contagion probability models from Chap. 2 that are used to generate random voting situations on the six possible linear voter preference rankings on three candidates in Fig. 4.1.

Recall that the P-E urn model for our particular case of interest starts out with six balls of different colors in an urn, with each color representing one of the preference rankings in Fig. 4.1. A ball is drawn at random and its color is noted to reflect the preference ranking for the first voter. That ball is then replaced in the urn with

Table 4.1 Limiting Condorcet Efficiency values for PR, NPR and BR with IC and IAC

Voting rule	IC	IAC
PR	.7572	.8815
NPR	.7572	.6296
BR	.9012	.9111

Fig. 4.1 The six possible linear preference rankings on three candidates

A	A	B	C	B	C
B	C	A	A	C	B
C	B	C	B	A	A
n_1	n_2	n_3	n_4	n_5	n_6

Table 4.2 Condorcet Efficiency values for P-E models with $n = 31$

k	PR	BR	Max efficiency	Max range	Same ranking
0	.7877	.9069	.9103	$7/15 < \lambda < 8/17$	1/6 = .1667
1	.8826	.9118	.9269	$7/19 < \lambda < 3/8$	2/7 = .2857
2	.9203	.9202	.9400	$7/24 < \lambda < 5/17$	3/8 = .3750
3	.9417	.9276	.9505	$5/21 < \lambda < 1/4$	4/9 = .4444
4	.9553	.9340	.9601	$1/17 < \lambda < 1/16$	5/10 = .5000

k additional balls of the same color as the ball that was just drawn, and the process then continues sequentially to obtain preference rankings for all n voters, with k additional balls being added at each step that are the same color as the ball that was just drawn in that step.

Table 4.2 lists results from Gehrlein (2003) for the case of $n = 31$ voters for P-E Models with each $k = 0, 1, 2, 3, 4$. These results list the Condorcet Efficiency of both PR and BR, along with the maximum value of efficiency that is obtained with any $WSR(\lambda)$ and the range of λ values that yield that maximum efficiency.

The Condorcet Efficiency results in Table 4.2 follow the pattern that has already been generally observed for the transition from IC with $k = 0$ to IAC with $k = 1$ for the limiting case for voters. Beyond that, as k is increased to introduce greater degrees of dependence among voters' preferences, the efficiency of BR continues to rise slowly. The corresponding efficiencies of PR have a much faster rate of increase and PR actually overtakes the efficiency of BR at $k = 2$, but the dominance of PR over BR remains small for $k \geq 2$. The $WSR(\lambda)$ range that achieves maximum efficiency shifts dramatically from a range that surrounds BR with $k = 0$ toward much smaller ranges that are close to PR for larger k. However, it is important to note that the differences between BR, PR and the most efficient $WSR(\lambda)$ are quite small for all $k \geq 2$.

These observations all make sense when the degree of dependence that is introduced as k increases is considered. In previous analysis, we considered this degree of dependence in terms of the likelihood that the first two balls that are drawn in a P-E Model based urn experiment will have the same color, so that the first two voters will have identical preference rankings. This is shown in Table 4.2 for each $k = 0, 1, 2, 3, 4$, where we see for example that when $k = 4$ the second draw from the urn will be taken at random from a total of $6 + 4 = 10$ balls with $1 + 4 = 5$ that are the same color as the first ball that was drawn. There is therefore a 50% chance that the second voter will have exactly the same preference ranking as the first voter. That is a lot of dependence! And, this will most likely lead to scenarios with voting situations with most voters having preferences that are clustered around only one or two of the possible preference rankings. Lepelley et al. (2000) perform a similar analysis with 101 voters for P-E based experiments, to find that as k gets at all large, random voting situations become very likely to have some candidate that is ranked as most preferred by a majority of voters, so that it must be both the CW and the winner by PR. Increased dependence among voters' preferences is clearly very agreeable with the use of PR.

By restricting attention to $k \leq 1$, we are focusing on cases where the selection of a voting rule is much more likely to have some possible impact on the outcome of an election. Even with $k = 1$, the differences in the overall IAC Condorcet Efficiency values for PR, BR and the most efficient $WSR(\lambda)$ are not very dramatic. However, since we are not anywhere near the end of this chapter, it obviously follows that we are not yet ready simply to conclude that PR works quite well and that it is not worth it to require voters to rank candidates so that some $WSR(\lambda)$ like BR can be used.

Given everything that has been observed so far, the analysis that follows will not consider any $WSR(\lambda)$ other than BR, due to the previously stated necessity to consider the transparency and complexity of voting rules. As a hypothetical scenario that illustrates this concern, suppose that a group has gathered to elect the winner in a three-candidate election. The first decision for the group is to determine how the election will take place. Somebody initially makes the obvious suggestion of using PR, so that the winner can be determined very easily and quickly by a sequential show of hands by voters in favor of each candidate. That option is as easy as it can be. But, somebody else then suggests that they want the voting to be done anonymously, so that it should be done by secret ballot, with the ballots subsequently being quickly counted in front of the entire group. So far, no significant resistance has been raised. Then a third person suggests that since ballots are being submitted anyway, the voters might have more input by providing preference rankings on the three candidates. Some members of the group now start to express concerns about what weights will be assigned to the candidate ranking positions to determine the winner, what trusted person will be responsible to honestly perform the tally from the ballots and how long it will take to finally determine the winner. If it is concluded that rankings should be used, the obvious natural sounding assignment of weights to rankings as 3:2:1 would probably be suggested immediately, exactly like Borda's Election by Order of Merit. This would typically be viewed as being logical to the members of the group, who are now content that an acceptable election procedure has apparently finally been established.

Then, another person in the back of the room stands to speak and correctly informs this group that there is probably some degree of dependence among the voters' preference rankings, as represented by something like IAC, and that a study by Cervone et al. (2005) indicates that the maximum Condorcet Efficiency of a $WSR(\lambda)$ with a large electorate occurs with the quite non-intuitive $\lambda \approx .3723$ in that case, so that the natural sounding 3:2:1 weights that have been suggested will not maximize Condorcet Efficiency. It is then added that this observation can be fully explained to the group if several hours of time would be allocated to do so.

After the meeting is finally brought back under some semblance of control by the chairperson, and this person in the back of the room has been ordered to remain silent for the remainder of the meeting, what should this person do? The additional knowledge that this maximizing $WSR(\lambda)$ does not yield any dramatic improvement in Condorcet Efficiency compared to BR anyway should lead this person to the wise decision to sit and remain silent, to thereby avoid the distinct possibility of being physically ejected from the meeting room. It is very important that the electorate has some natural understanding of, and confidence in, the logic of the process that is being used to determine the winner in an election. With that principle in mind, BR is used as the sole basis for evaluating all $WSR(\lambda)$, except for the obvious cases of PR and NPR that do not require rankings.

4.4 Efficiency Relationships to Group Mutual Coherence

While the overall Condorcet Efficiency values for PR and BR do indeed become closer as dependence is introduced among the voters' preferences with IAC, it is of interest to consider what happens as values of the parameters that measure Group Mutual Coherence are changed. Changes in these parameters had a significant impact on the probability that Condorcet's Paradox is observed, and the Condorcet Efficiencies of voting rules will also behave differently as the values of different parameters change, and some of these patterns of change are obvious and very predictable. Consider for example the case in which it is known that there is a strong reason to believe that a Perfect Strong Positively Unifying Candidate nearly exists, so that the proportion of voters $\frac{t^*}{n}$ will be large, since some candidate is frequently ranked as being most preferred. If $t^* > \frac{n}{2}$, then some candidate must be ranked as most preferred by more than half of the voters, and that candidate obviously must be both the CW and the winner by PR with a Condorcet Efficiency of one. As we move toward lower parameter values within the range $\frac{n}{3} \leq t^* \leq \frac{n}{2}$, the Condorcet Efficiency of PR can then be expected to consistently decrease with smaller values of t^*.

The link between some other parameters and the Condorcet Efficiency of voting rules is not always as obvious as that case. For example, similar observations can be made for Weak Measures, where it might seem odd to see at first glance that the Condorcet Efficiency of PR is one if $t = 0$, so that a perfect Weak Negatively Unifying Candidate exists, and some candidate is never ranked as most preferred by any voter. But, the remaining two candidates must then split the top-ranked positions between them. One of these two candidates must therefore be ranked as most preferred by more than half of any odd number of voters, and therefore $t^* > \frac{n}{2}$ whenever $t = 0$. The Condorcet Efficiency of PR can then generally be expected to consistently decrease as t gets larger.

4.4.1 *Condorcet Efficiency Relationships with Weak Measures*

A theoretically-based analysis of the relationships between the Condorcet Efficiency of voting rules and the parameters of Group Mutual Coherence is performed in Gehrlein et al. (2011), which uses the notions behind some of the definitions from the preceding chapter. That is, a representation is first obtained for the limiting Condorcet Efficiency as $n \to \infty$ for PR, NPR and BR as a function of the proportion of voters α_b who must have their preferences ignored before a Perfectly Weak Positively Unifying Candidate exists. Then, $CE_{VR}^{WPU}(\infty, IAC(\alpha_b))$ denotes this Condorcet Efficiency with the $IAC(\alpha_b)$ assumption for voting rule VR. The results are displayed graphically in Fig. 4.2, where it is observed that BR dominates PR, which in turn dominates NPR over the entire range $0 \leq \alpha_b \leq \frac{1}{3}$. The dominance of BR over PR is not highly significant, but the performance of NPR is consistently very weak. This observation is totally in conformance with the expected overall Condorcet Efficiency results that were listed in Table 4.1.

Based on the discussion above, PR must start out very well and have an efficiency value of one for $\alpha_t = 0$, and the graphed values for $CE_{VR}^{WNU}(\infty, IAC(\alpha_t))$ that are displayed in Fig. 4.3 verify that this is true, and show that PR actually outperforms BR over the range $0 \leq \alpha_t \leq .15$. Unfortunately, there are extremely disappointing results that follow for PR as α_t increases. PR only has a Condorcet Efficiency of about .33 as $\alpha_t \to \frac{1}{3}$, so it is not performing any better than a random selection process for determining the winner. The results in Fig. 4.3 again consistently show generally poor performance for NPR.

The Efficiency results in Fig. 4.4 for $CE_{VR}^{Pol}(\infty, IAC(\alpha_c))$ are similar to those in Fig. 4.2, with BR dominating PR, which in turn dominates NPR over the entire range $0 \leq \alpha_c \leq \frac{1}{3}$. The dominance of BR over PR is again not very significant, but very poor performance is again displayed by NPR.

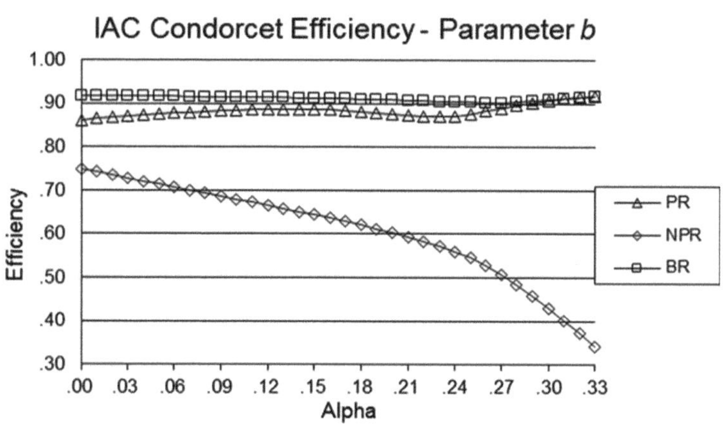

Fig. 4.2 Limiting IAC Condorcet Efficiency of PR, NPR and BR with parameter *b*

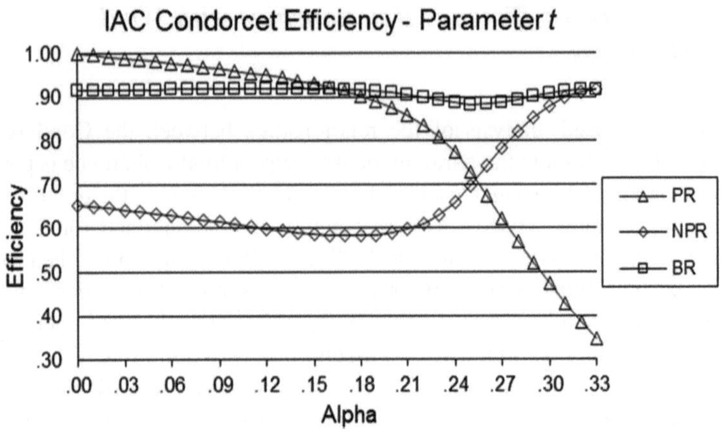

Fig. 4.3 Limiting IAC Condorcet Efficiency of PR, NPR and BR with parameter *t*

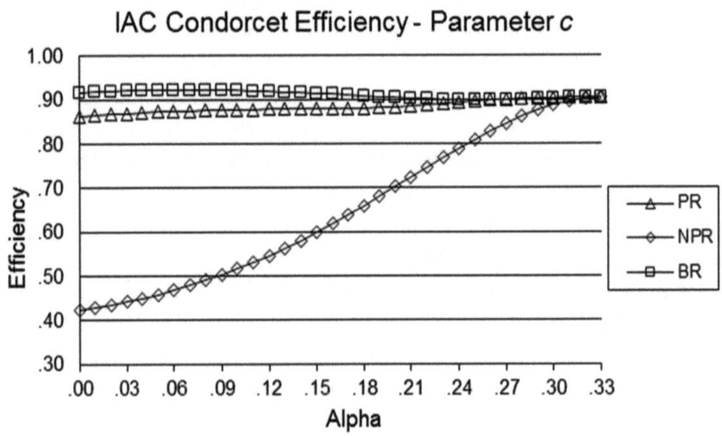

Fig. 4.4 Limiting IAC Condorcet Efficiency of PR, NPR and BR with parameter *c*

The general conclusion to be reached from these theoretical results with Weak Measures is that PR only shows greater levels of Condorcet Efficiency than BR for values of Parameter *t* in the range $0 \leq \alpha_t \leq .15$, and there is not a particularly significant difference in that case. The critical issue is that there are identifiable scenarios for which PR displays terrible performance, and this never happens for BR. The efficiency of NPR is generally so bad that it does not appear to be a viable option for consideration. By using BR, we are usually, but not always, obtaining the best result. However, when PR does beat BR there is not much of a difference, and PR has scenarios in which it has terrible performance while BR never does that. It is starting to sound like BR just might generally be a good choice that also avoids the possibility of having a really bad outcome.

Obvious questions remain as to whether or not this is what actually happens in reality. An empirically-based analysis by Gehrlein and Plassmann (2014) used the Tideman-Plassmann Simulation Model to answer these questions. First, does PR really exhibit such terrible performance as $\alpha_t \rightarrow \frac{1}{3}$? Figure 4.5 shows simulated estimates of the Condorcet Efficiency of PR with one million voters for Parameter t with each of the databases ANES, PB and ERS, along with the associated theoretical limiting IAC values.

Two things are very evident from Fig. 4.5. The most obvious fact is that PR really does show terrible performance as $\alpha_t \rightarrow \frac{1}{3}$ for all databases. It is also evident that the empirically-based results generate higher levels of efficiency for PR than what is suggested by IAC. However, the critical issue is that all of these results show exactly the same pattern of behavior as α_t changes, and PR really can be expected to result in very poor performance under these conditions.

The second question asks, does NPR really exhibit such overall poor performance as $\alpha_b \rightarrow \frac{1}{3}$? Figure 4.6 shows the simulated Condorcet Efficiency estimates for NPR with the three databases and IAC.

These results are much the same as those from the analysis of PR. There is more of a spread between the curves and all empirically-based results have greater efficiency than IAC, but they all display the same pattern of change and they all show terrible performance for NPR as $\alpha_b \rightarrow \frac{1}{3}$.

We have found that empirically-based simulation estimates for the efficiency of both PR and NPR really do exhibit the very poor expectations from our IAC analysis. The remaining question is: Does BR really consistently perform very well in all scenarios? This is answered in part in Fig. 4.7 which shows the simulated Condorcet Efficiency estimates for BR with the three databases and IAC for Parameter t. The answer is definitely "Yes." The simulated efficiency estimates from all databases slightly exceed the already high values from the analysis with IAC.

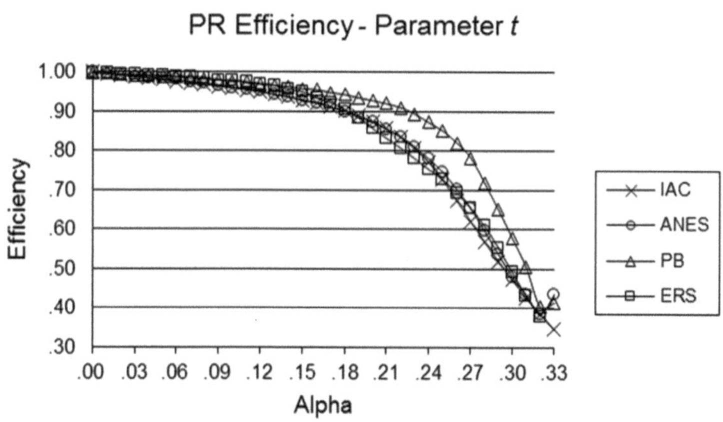

Fig. 4.5 Empirically-based efficiency estimates for PR with parameter t

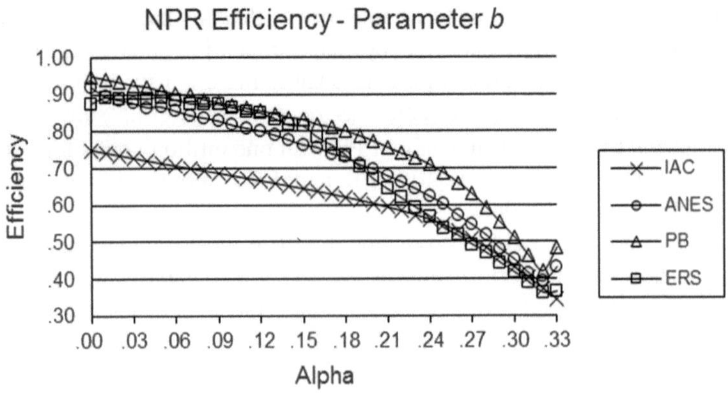

Fig. 4.6 Empirically-based efficiency estimates for NPR with parameter *b*

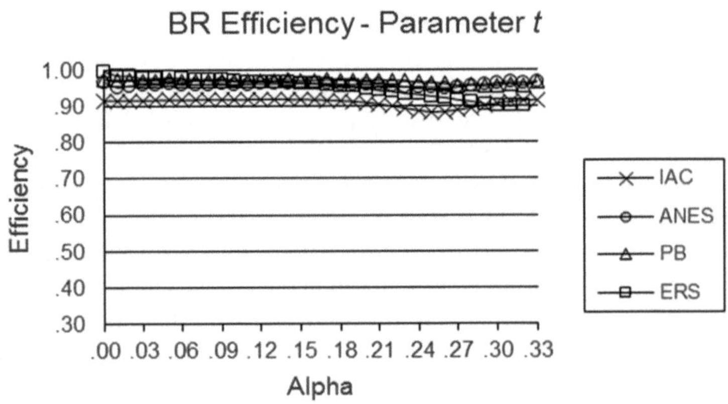

Fig. 4.7 Empirically-based efficiency estimates for BR with parameter *t*

The graphs of efficiency values for BR are not shown for Parameters *b* and *c*, since they are nearly identical to what is observed in Fig. 4.7 for Parameter *t*. So, BR really does perform well over the entire ranges of parameters for all three Weak Measures.

4.4.2 Condorcet Efficiency Relationships with Strong Measures

The next step in the analysis of Gehrlein et al. (2011) was to consider the relationships between Condorcet Efficiency and the parameters of Strong Measures for PR, NPR and BR, to see if same general conclusions from the analysis of Weak

Measures remain valid. Figure 4.8 shows plots of $CE_{VR}^{SNU}(\infty, IAC(\alpha_{b^*}))$ for PR, NPR and BR.

The efficiency values in Fig. 4.8 again show BR dominating PR, which in turn dominates NPR. The margin of dominance is small for BR over PR, and the performance of NPR is very poor over the entire range $\frac{1}{3} \leq \alpha_{b^*} \leq 1$.

Condorcet Efficiency values for $CE_{VR}^{SPU}(\infty, IAC(\alpha_{t^*}))$ are shown in Fig. 4.9, where we know from previous discussion that PR has a Condorcet Efficiency of one whenever $\frac{1}{2} \leq \alpha_{t^*} \leq 1$.

Just as in the case with Parameter t, PR does dominate BR over a small range of α_{t^*}, but for decreasing values with $\alpha_{t^*} \rightarrow \frac{1}{3}$ we find again that PR behaves almost like a random procedure for selecting a winner. NPR generally performs poorly, but it still has greater efficiency than PR for smaller values of α_{t^*}.

NPR is not going to have any possible chance of being considered as a viable option when it is analyzed with Parameter c^*, given the following observation. If $c^* = n$ in a voting situation, some candidate is always middle-ranked and that candidate obviously will always be a winner with NPR. But, if that candidate is always middle-ranked, it follows that $t = 0$ for that voting situation. Based on the earlier analysis of such cases, some other candidate must therefore be the CW. So, the Condorcet Efficiency of NPR must be zero in this case, and remain very low for relatively large values of c^*. This is clearly observed in the limiting $CE_{NPR}^{Cen}(\infty, IAC(\alpha_{c^*}))$ efficiencies that are shown in Fig. 4.10.

BR again dominates PR over the entire range $\frac{1}{3} \leq \alpha_{c^*} \leq 1$ in Fig. 4.10. Overall, the conclusions with Strong Measures are completely in agreement with those that were drawn with Weak Measures.

Empirically-based results from Gehrlein and Plassmann (2014) also verify that PR does indeed tend to behave like a random selection process for picking winners when $\alpha_{t^*} \rightarrow \frac{1}{3}$ in Fig. 4.11.

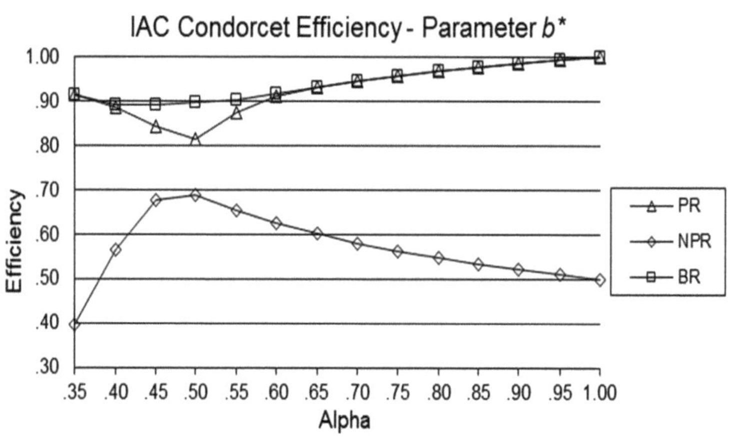

Fig. 4.8 Limiting IAC Condorcet Efficiency of PR, NPR and BR with parameter b^*

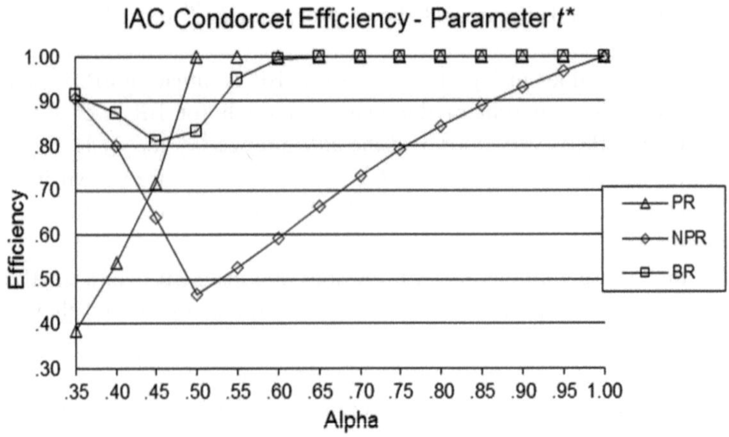

Fig. 4.9 Limiting IAC Condorcet Efficiency of PR, NPR and BR with parameter t^*

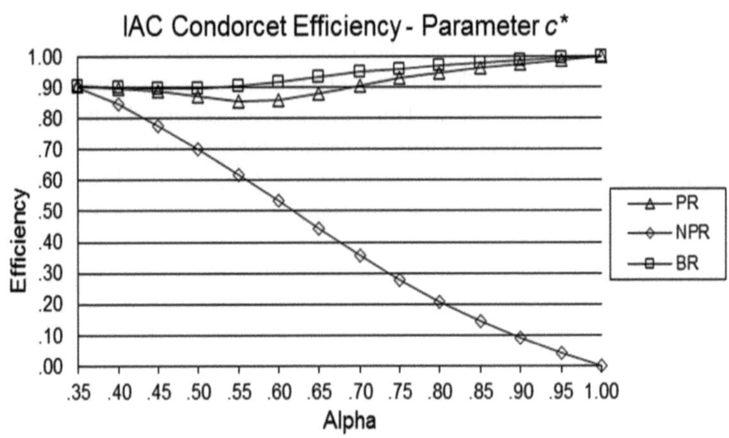

Fig. 4.10 Limiting IAC Condorcet Efficiency of PR, NPR and BR with parameter c^*

There is no need to verify the observation that NPR has disastrous performance as $\alpha_{c*} \rightarrow 1$, given the discussion that was presented earlier. The fact that BR performs well in all scenarios is partially verified by the empirically-based simulation estimates in Fig. 4.12 for the worst case BR performance that was observed with Parameter t^* in Fig. 4.9.

These empirically-based efficiencies for BR are all improved from the result with IAC for Parameter t^*, to give it even stronger support than the theoretical results suggested. The simulated BR efficiency graphs for Parameters b^* and c^* display identical behavior, so they are not included.

We have therefore consistently seen for both Weak and Strong Measures that BR typically has the greatest Condorcet Efficiency, but there are a few instances in which PR does perform marginally better than BR. However, PR has identifiable

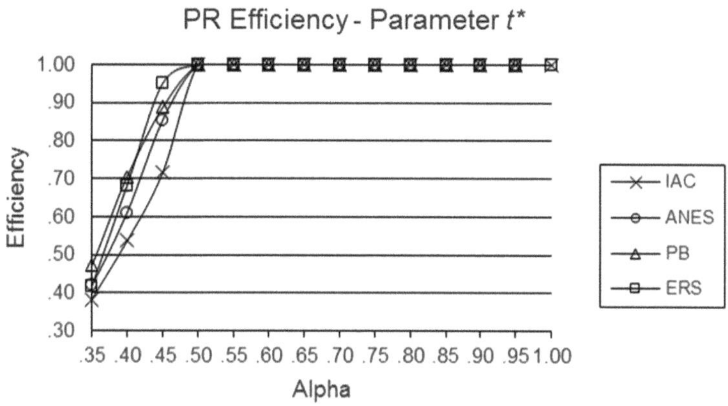

Fig. 4.11 Empirically-based Efficiency estimates for PR with parameter t^*

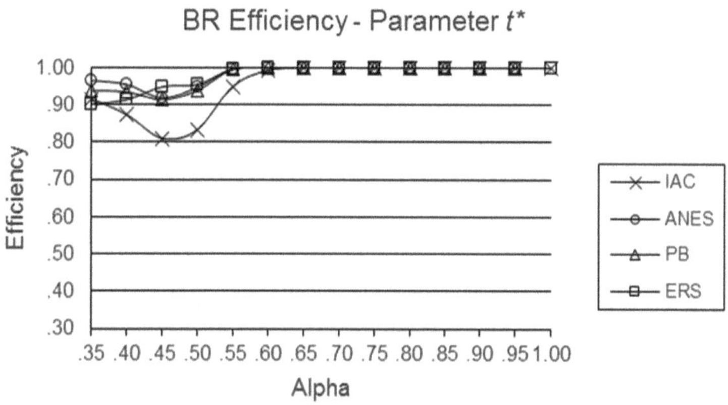

Fig. 4.12 Empirically-based Efficiency estimates for BR with parameter t^*

regions in which its performance is similar to that of a random selection process. NPR consistently shows very poor performance. This led Gehrlein et al. (2011) to suggest the *Borda Compromise* that asserts that BR should be used to completely eliminate the possibility that very poor performance could occur.

Before we completely accept the Borda Compromise, there is an additional issue to consider. It is important to obtain some measure of the likelihood that taking the step to require rankings from voters so that BR can be implemented will actually have an impact on the result of the election. That is, how likely is it that we will get the same election outcome by using BR that we would have obtained simply by using PR or NPR?

This idea has been considered by Moulin (1988) and Saari (1990) who show that whenever both PR and NPR elect the same winner in a three-candidate election, then every $WSR(\lambda)$, including BR, must elect that same candidate. Table 4.3 shows

Table 4.3 Probability that all $WSR(\lambda)$ elect the same candidate with $n = 101$

Base	Same winner
IAC	.5082
ANES	.5923
PB	.7433
ERS	.7811

probabilities that PR and NPR will elect the same winner with 101 voters from Gehrlein et al. (2016a).

The probability that all $WSR(\lambda)$ elect the same winner ranges from .5082 to .7811, so the chance that the selection of a single-stage voting rule might actually have an impact on the outcome ranges from .2189 to .4918. This indicates that there can be a reasonably good likelihood that the choice of a single-stage voting rule could have a real impact on the outcome of an election, to show that the Borda Compromise is meaningful in a significant proportion of cases.

It is important to note that the results in Table 4.3 indicate that there are some significant differences between the levels of underlying internal structure that exists among the preferences of the voters in these different databases. This explains the differences in the Condorcet Efficiency estimates that have been observed across the three empirically-based databases that have been used to calibrate the simulation models. These differences could result from the level of knowledge that voters might have about the candidates, the level of significance that voters place on the outcome of the elections and other factors that were discussed in the immediately preceding chapter. The simple metrics that have been used to measure Group Mutual Coherence clearly are not able to completely react to all of the intricacies of the probably more complex models of Group Mutual Coherence that govern some of these scenarios. However, it is very important to note that despite their simplicity, these basic metrics have still proven to be totally adequate to typically produce consistent patterns for the Condorcet Efficiency values over their ranges for all databases.

4.5 Modifications to Restrict Attention to Critical Situations

There is definitely a non-trivial proportion of voting situations for which all $WSR(\lambda)$ elect the same winner over all databases in Table 4.3. Since none of these voting situations have any impact whatsoever on the determination of which single-stage voting rule has the greatest Condorcet Efficiency, Diss and Gehrlein (2015) suggest that they should be ignored in any comparative evaluation of single-stage voting rules. This is pursued by developing the *Modified IAC Assumption* (MIAC) for which all voting situations such that PR and NPR do not elect the same winner are equally likely to be observed.

Table 4.4 Limiting Condorcet Efficiency of $WSR(\lambda)$ with IAC and MIAC

λ	Condorcet Efficiency	
	IAC	MIAC
.00	.8815	.7668
.10	.8979	.8027
.20	.9123	.8342
.30	.9227	.8567
.40	.9249	.8617
.50	.9111	.8315
.60	.8720	.7461
.70	.8176	.6273
.80	.7560	.4926
.90	.6919	.3528
1.00	.6296	.2167

The use of MIAC does not have the same magnitude of impact on the Condorcet Efficiency for all voting rules, which is clearly shown in Table 4.4 from Diss and Gehrlein (2015) that lists the limiting Condorcet Efficiency of various $WSR(\lambda)$ as $n \rightarrow \infty$ with both IAC and MIAC.

The maximum amount of reduction in efficiencies from using MIAC for these $WSR(\lambda)$ occurs at the extremes of PR and NPR, and it is minimized around the center of the range near BR. The immediate conclusion is that despite having reduced efficiencies for all voting rules, BR efficiency looks much better on an overall basis, compared to both PR and NPR, when attention is restricted to the cases where the selection of a single-stage voting rule might have an impact on the election outcome. This tells us what happens on an overall basis when MIAC is used, but the impact of this modification could be very uneven across different regions of possible parameter values that measure Group Mutual Coherence.

4.5.1 *Modifications to the Databases: Weak Measures*

The impact that the MIAC restriction has on the Condorcet Efficiency of PR, NPR and BR as parameter values are changed was considered by Gehrlein et al. (2016b), by using the same general process of analysis that was utilized above. For example, $CE_{VR}^{WPU}(\infty, MIAC(\alpha_b))$ represents the limiting Condorcet Efficiency of voting rule VR as $n \rightarrow \infty$ under the MIAC assumption when the minimum proportion of all voters that must have their preferences ignored before a Perfectly Weak Positively Unifying candidate exists in the reduced voting situation is α_b. The theoretical MIAC-based efficiency results that are obtained for Parameters b and c look much the same as those that are given for IAC in Figs. 4.2 and 4.4 respectively. That is, BR dominates PR, which in turn dominates NPR over the entire range of parameter values. The difference that occurs with MIAC is that all of the corresponding

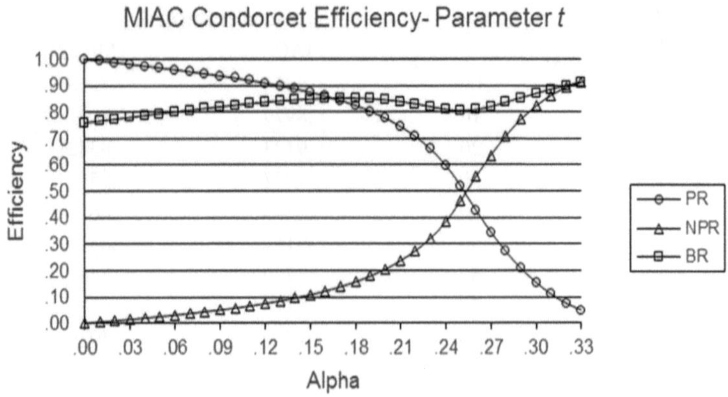

Fig. 4.13 Limiting MIAC Condorcet Efficiency of PR, NPR and BR with parameter *t*

efficiency values are reduced, and the efficiency differences between voting rules tend to be more pronounced.

Some very striking results are observed with MIAC for Parameter *t*, as shown in Fig. 4.13.

The MIAC efficiency results for Parameter *t* in Fig. 4.13 epitomize the logic behind the Borda Compromise. We see that PR does start out very well and it actually outperforms BR on the basis of Condorcet Efficiency over the range $0 \leq \alpha_t \leq .15$, just like it did with IAC in Fig. 4.3. The efficiencies are smaller with MIAC, but the margin of dominance is slightly increased. As α_t increases for values above that range, BR efficiency remains relatively stable, while the efficiency of PR plummets to near zero levels for large values of α_t. PR now exhibits even worse performance than it did with IAC, where it acted like a random chooser of a winner as $\alpha_t \rightarrow \frac{1}{3}$. NPR also acted like the very poor performer with IAC, and it now becomes a terrible performer with MIAC to have near zero efficiency over a significant range of α_t values. NPR efficiency does increase dramatically for larger values of α_t, but it never dominates BR. Much more dramatic results are observed with MIAC than with IAC, and the consistent theme remains. While BR does not always have the highest level of efficiency over the entire range for all parameters of the Weak Measures, it never performs poorly; while scenarios exist for which both PR and NPR could be performing at horribly low levels of efficiency.

It is of course necessary to provide some evidence that this potentially horrible performance by PR and NPR is a real issue, and Gehrlein et al. (2016b) provides this evidence with empirically-based simulations from the Tideman-Plassmann Model. Figure 4.14 shows the results for PR with the modified databases MANES, MPB and MERS, along with the theoretical MIAC results.

All databases in Fig. 4.14 display the same pattern of decreasing efficiency for PR from one to zero as α_t increases over the range $0 \leq \alpha_t \leq \frac{1}{3}$, so PR really does display extraordinarily poor performance for larger values of α_t. By comparing these results to Fig. 4.5, the efficiency values for PR with Parameter *t* definitely tend

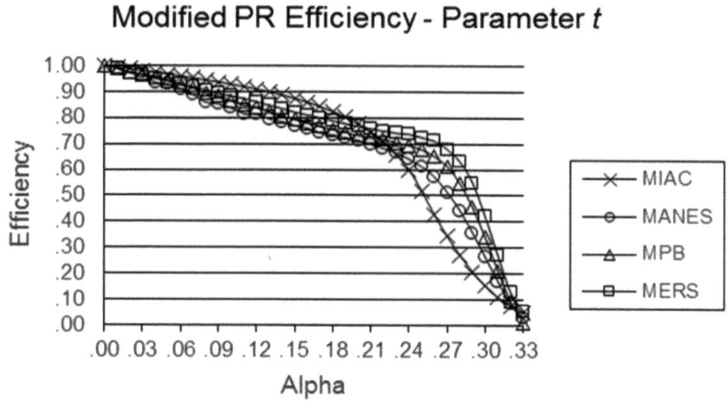

Fig. 4.14 Empirically-based modified Efficiency estimates for PR with parameter t

to be more widely distributed with the modified databases. By removing the voting situations for which all $WSR(\lambda)$ elect the same winner, the scenarios that are displaying high degrees of concordance are being eliminated. This leads to having remaining databases without as much strong internal structure, so that the remaining voting situations in each modified database might be more susceptible to being controlled by its own particular intricacies.

The same general conclusions are drawn for the Condorcet Efficiency of NPR with the modified databases for Parameter t in Fig. 4.15.

Empirically-based simulation results for the Condorcet Efficiency of BR with the modified databases for Parameter t are shown in Fig. 4.16.

The BR efficiencies for the modified databases are reduced in Fig. 4.16 from their associated values in Fig. 4.7, but BR never has the same poor performance as PR and NPR. The overall impact of other structural differences that are embedded within the different databases are definitely evident in these results, since the same shifts in BR efficiency are not observed in all of the transitions to the modified databases. That is, the empirically-based results in Fig. 4.16 are not consistent with what we observed with IAC in Fig. 4.7, since they are not always showing greater efficiency for BR than the theoretical MIAC results. This is particularly noticeable in the MANES database. However, the general patterns of behavior do in fact remain intact relative to our simple measures of Group Mutual Coherence across the modified databases.

4.5.2 Modifications to the Databases: Strong Measures

Computed limiting Condorcet Efficiency values as $n \to \infty$ with MIAC from Gehrlein et al. (2016b) continue to show that BR dominates PR, which in turn dominates NPR for Parameters b^* and c^* over the entire range of parameter values.

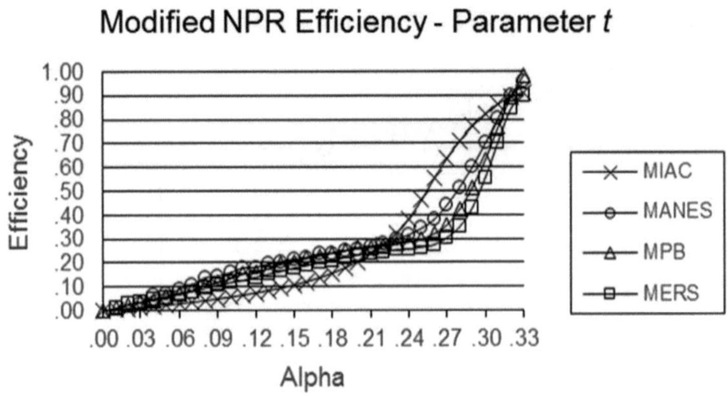

Fig. 4.15 Empirically-based modified Efficiency estimates for NPR with parameter t

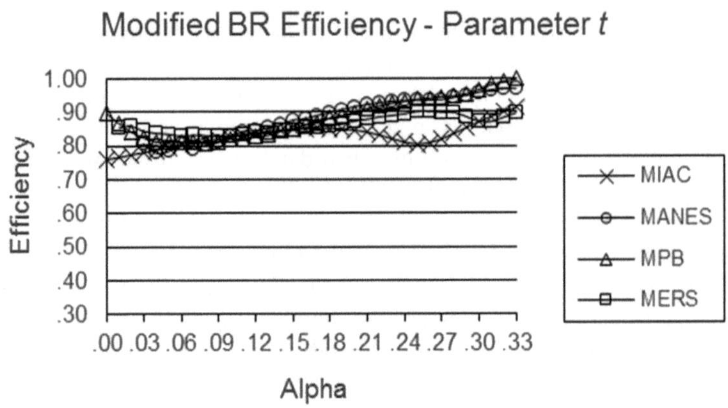

Fig. 4.16 Empirically-based modified Efficiency estimates for BR with parameter t

The only occurrence for which PR has greater efficiency than BR exists for Parameter t^*, as seen in Fig. 4.17.

The Condorcet Efficiency of PR was found to be equal to one earlier for $\frac{1}{2} \leq \alpha_{t^*} \leq 1$ with IAC, and it is not reduced in that range for MIAC. This follows from the fact that this result was shown to be true simply because $\frac{1}{2} \leq \alpha_{t^*} \leq 1$, regardless of how that result occurred. There is however a dramatic change that results from the transition from IAC to MIAC. If PR elects the CW in a voting situation that is allowable with MIAC, it then follows from definition that NPR cannot also elect that same winner, so that the Efficiency of NPR must therefore be zero for $\frac{1}{2} \leq \alpha_{t^*} \leq 1$, as observed in Fig. 4.17.

Similar to earlier observations for Parameter t with MIAC in Fig. 4.13, Parameter t^* has regions in which both PR and NPR have Condorcet Efficiency values near zero. PR does have greater efficiency values than BR for the range $.48 \leq \alpha \leq .61$, but the BR efficiency remains above .67 over the entire range of α_{t^*}. While BR does

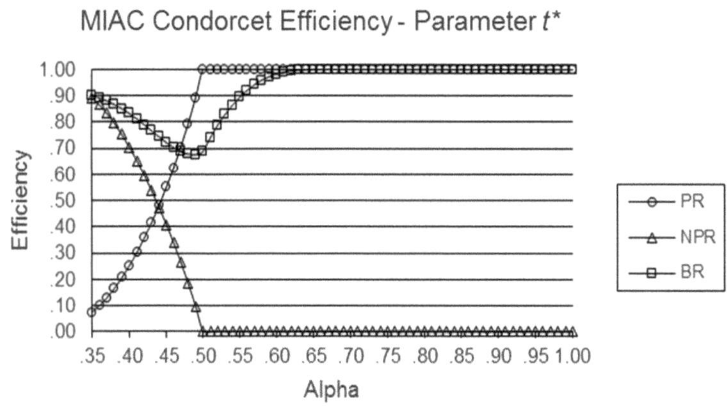

Fig. 4.17 Limiting MIAC Condorcet Efficiency of PR, NPR and BR with parameter t^*

not always have the greatest values for efficiency, it avoids the potential of having significant problems of near zero efficiency with both PR and NPR.

These conclusions were backed up by empirically-based simulations with the three databases. The results for Parameters b^* and c^* with MIAC all showed BR dominating PR, which in turn dominated NPR. The plots of the PR efficiency for Parameter t^* are given in Fig. 4.18, to verify that PR does indeed have a near zero efficiency for $\alpha_{t^*} \to \frac{1}{3}$.

There is no need to verify the terrible performance of NPR, since its zero efficiency with MIAC must always hold whenever $\frac{1}{2} \leq \alpha_{t^*} \leq 1$. The plotted efficiencies for BR with Parameter $t^*$are given in Fig. 4.19, where we see again that the empirically-based simulation results produce greater efficiency estimates than those obtained with MIAC. This strongly supports the conclusion that BR maintains relatively good performance over the entire range of α_{t^*}, while PR and NPR both collapse in different parts of this same range of parameter values.

Two factors consistently arise as a result of using these modified databases. The first is that the Condorcet Efficiency values are reduced, except in cases where they cannot be smaller, such as when the Condorcet Efficiency of PR is one when $\frac{1}{2} \leq \alpha_{t^*} \leq 1$. In addition, the previously mentioned intricacies of possibly more complex models of Group Mutual Coherence that might be embedded within some of these empirical databases have a greater impact on the modified databases, to cause unequal shifts in efficiency values across the databases.

The modified databases yield even larger regions of poor performance for PR and NPR than the original analysis, while less dramatic changes are observed in the Condorcet Efficiencies for BR. This all strongly reinforces the reasoning behind the Borda Compromise when attention is restricted to voting situations in which the selection of a single-stage voting rule could have an impact on the election outcome.

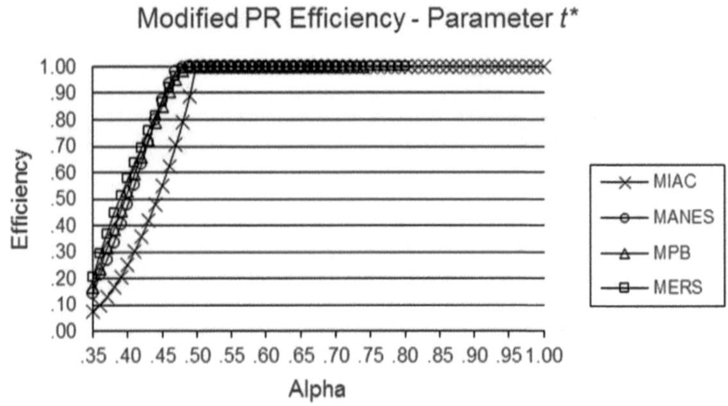

Fig. 4.18 Empirically-based modified Efficiency estimates for PR with parameter t^*

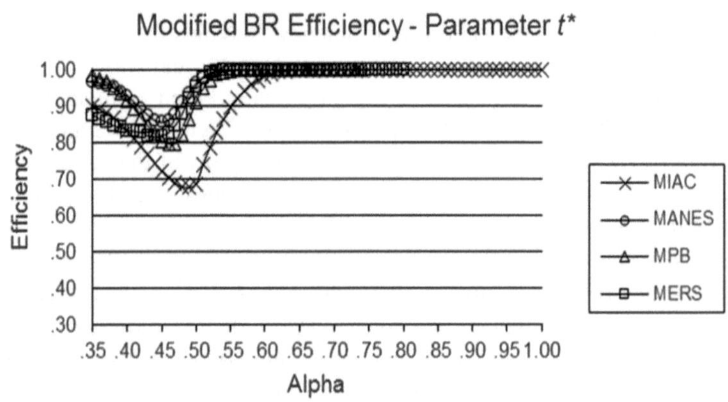

Fig. 4.19 Empirically-based modified Efficiency estimates for BR with parameter t^*

4.6 Refined Measures of Group Mutual Coherence

The discussion of Refined Measures of Group Mutual Coherence in the preceding chapter showed that significantly different results are obtained for the limiting probability that Condorcet's Paradox is observed, depending upon which parameter is the minimum of b, t and c, or which one is the maximum of b^*, t^* and c^*. Much stronger relationships were also found to exist between those refined parameter values and the probability that Condorcet's Paradox will be observed, relative to what was found with the original parameter definitions. The probability that a CW exists is directly linked to Condorcet Efficiency, so Gehrlein and Lepelley (2016) extended the same general process of analysis that has been used above to consider the Condorcet Efficiency of voting rules as a function of the Refined Measures of Group Mutual Coherence.

Table 4.5 Limiting IAC efficiencies for PR, NPR and BR with refined weak measures

Voting rule	Minimum parameter		
	b	t	c
PR	.8489	.9451	.8480
NPR	.7561	.6640	.4561
BR	.9065	.9065	.9210

The analysis starts on an overall basis and determines the limiting IAC Condorcet Efficiency of PR, NPR and BR as $n \to \infty$ when attention is restricted to voting situations in which each of the Weak Measures b, t and c is the minimum of the three. The results are summarized in Table 4.5.

It was noted in the preceding chapter that the probability of observing Condorcet's Paradox is not the same when each of the Weak Measures is the minimum, and it is not surprising that the same outcome occurs for Condorcet Efficiencies. There are distinct differences in performance for the voting rules in the different scenarios for the minimum parameter values. The greatest efficiency occurs for PR when t is the minimum. But, we have already learned that having a very small value for t will very probably lead to $t^* > \frac{n}{2}$ to ensure that PR must elect the CW. It is no longer surprising to see that NPR is the worst performer in all three scenarios and that BR has very stable performance for all three cases. The very high efficiency of PR when t is the minimum suggests that the trends in PR efficiency values in Fig. 4.3 might be significantly modified when attention is restricted to those particular voting situations.

The corresponding efficiencies for the Strong Measures for voting situations in which each of b^*, t^* and c^* is the maximum of the three parameters are listed in Table 4.6, where some dramatic differences in results are observed.

Table 4.6 shows an amazing efficiency .9879 for PR when t^* is the maximum of b^*, t^* and c^*. The results also give a very big advantage to BR over both PR and NPR when c^* is the maximum of the three parameters. All of this suggests that there could be some significant shifts in the efficiency plots compared to what was observed with IAC, when attention is shifted to voting situations that account for which particular parameter is the maximum of the Strong Measures.

4.6.1 *Refined Parameters: Weak Measures*

Whenever we have $b = 0$ in a voting situation, it must also be true from its definition that $Rb = 0$, so the Condorcet Efficiency of any voting rule must be identical for both when $b = Rb = 0$. This fact must also obviously hold for both Parameters t and c. So, any differences that occur in the Condorcet Efficiencies of voting rules must be observed for larger values of these parameters as they increase from zero. Gehrlein and Lepelley (2016) obtained representations for limiting Condorcet Efficiencies of PR, NPR and BR with IAC as a function of parameter values for the Refined Weak Measures. Following earlier definitions,

Table 4.6 Limiting IAC efficiencies for PR, NPR and BR with refined strong measures

Voting rule	Maximum parameter		
	b^*	t^*	c^*
PR	.8675	.9879	.7778
NPR	.6819	.7901	.3910
BR	.8795	.9679	.8829

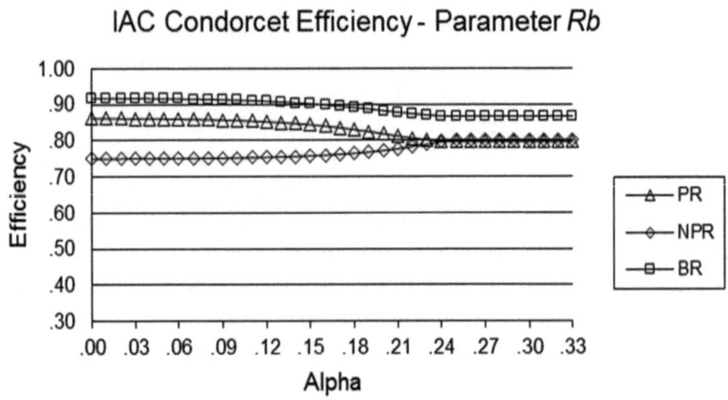

Fig. 4.20 Limiting IAC Condorcet Efficiency of PR, NPR and BR with parameter *Rb*

values of $CE_{VR}^{WPU}(\infty, IAC(\alpha_{Rb}))$ are plotted in Fig. 4.20 for PR, NPR and BR, to show the impact of the fact from Table 4.5 that both NPR performs at its overall best and that BR ties for its minimum efficiency when b is the minimum parameter that is observed. This turns out to have a very positive overall influence on the efficiencies of both of these voting rules.

By comparing the results with Parameter Rb in Fig. 4.20 to those in Fig. 4.2 for b, we observe significant changes in the behavior of efficiencies of voting rules as the modified parameter values increase from zero. The unsurprising result for BR is that it consistently dominates PR on the basis of Condorcet Efficiency in Fig. 4.20. The good news for BR is that it now does so with a significantly increased margin for larger values of α_{Rb} than we observed for larger values of α_b. The good news for NPR is that it no longer has efficiencies that plummet to behave like a random selection process as $\alpha_{Rb} \rightarrow \frac{1}{3}$. In fact, the NPR efficiencies are much improved with the revised parameter, but NPR still remains dominated by both PR and BR over the entire range of parameter values. The same results are observed for the behavior of PR and BR for Parameter c, making BR look substantially better by comparison to PR with the results for Rc.

It has already been noted that the use of Parameter Rt is expected to be beneficial to the relative performance of PR, and this outcome is clearly present in the plotted Condorcet Efficiency results in Fig. 4.21.

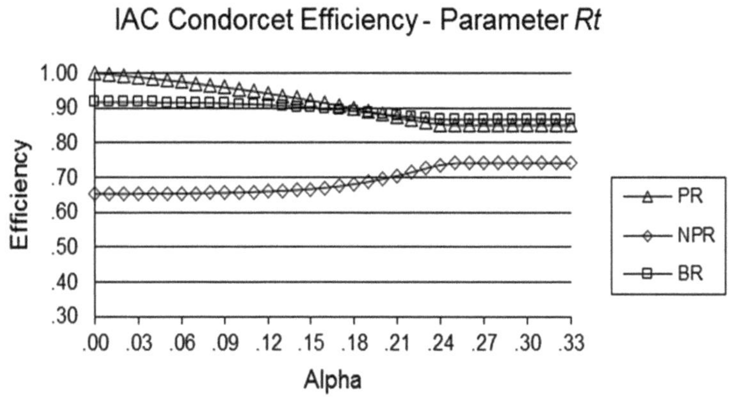

Fig. 4.21 Limiting IAC Condorcet Efficiency of PR, NPR and BR with parameter *Rt*

The precipitous drop in PR efficiency values for large values of α_t in Fig. 4.3 vanishes for α_{Rt} in Fig. 4.21. The NPR efficiencies remain quite similar, by consistently displaying levels of poor performance. The efficiency of PR improves dramatically with Parameter Rt, compared to Parameter t, since PR still performs better than BR for smaller values of α_{Rt}, as is did with α_t, but BR now is only showing marginally better efficiency than PR for larger values of α_{Rt}.

We have seen to this point that by focusing attention on the particular Weak Measure of Group Mutual Coherence that most appropriately fits the voters' preference rankings in a voting situation, significantly different Condorcet Efficiency results can be observed for PR, NPR and BR. However, BR still typically maintains an advantage over both of the other single-stage voting rules in most cases for Weak Measures. This advantage has been enhanced for Parameters Rb and Rc, but reduced for Rt.

4.6.2 *Refined Parameters: Strong Measures*

Since NPR has consistently performed so poorly, it cannot be considered to be a viable option, so we focus attention on comparing PR and BR. The results in Table 4.6 indicate that PR should have its best performance when Parameter t^* is the maximum of the three Strong Measures, and its worst performance when c^* is the maximum, so we evaluate those two cases. There is also a common starting point for Condorcet Efficiency values for any voting rule with Strong Measures, since if $t^* = n$, then it must also be true that $Rt^* = n$ by definition. The same relationship must also be true for both Parameters b^* and c^*. This requires for example that $CE_{VR}^{SPU}(\infty, IAC(\alpha_{t^*})) = CE_{VR}^{SPU}(\infty, IAC(\alpha_{Rt^*}))$ when $\alpha_{t^*} = \alpha_{Rt^*} = 1$

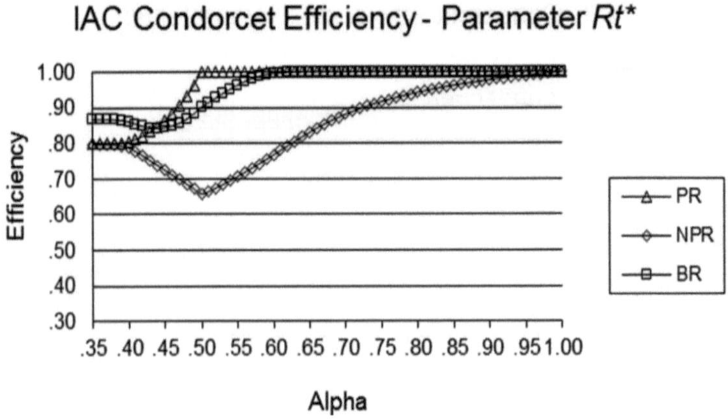

Fig. 4.22 Limiting IAC Condorcet Efficiency of PR, NPR and BR with parameter Rt^*

for any voting rule, so any Condorcet Efficiency differences that occur from using this Refined Strong Measure must occur for smaller values of α_{Rt^*}.

Representations for the limiting Condorcet Efficiency of $CE_{VR}^{SPU}(\infty, IAC(\alpha_{Rt^*}))$ are obtained in Gehrlein and Lepelley (2016) for PR, NPR and BR to consider what happens for the scenario that gives the strongest level of support for PR. The results are displayed in Fig. 4.22.

By comparing the PR Efficiency results in Fig. 4.22 to the associated results in Fig. 4.9, it is very clear that PR performs much better on the basis of the Refined Measure Rt^*, since its efficiency at $\alpha_{Rt^*} \to \frac{1}{3}$greatly exceeds the PR performance of acting like to a random chooser when $\alpha_{t^*} \to \frac{1}{3}$. This is all good news for PR, but what happens when we consider its expected worst case performance for the Refined Parameter Rc^* in Fig. 4.23?

The worst case scenario for PR in Fig. 4.23 sees its Condorcet Efficiency decrease significantly for smaller values of Rc^* relative to its performance with c^* in Fig. 4.10, where its performance was completely comparable to BR. While PR is not behaving like a random chooser here, its efficiency is only a little over .50 for smaller values of α_{Rc^*}. The performance of NPR is again horrible for this case, with an efficiency of zero when $\alpha_{Rc^*} \to 1$. Very similar patterns of behavior are seen for α_{Rb^*}, but neither PR nor NPR has efficiency values that are quite as low at the extreme points that we observe in Fig. 4.23.

The conclusions of the comparison of BR to PR and to NPR with Refined Measures of Group Mutual Coherence remain similar to, and generally supportive of, the previous observations that led to the Borda Compromise. The differences in the relative comparisons of efficiencies of PR and BR are however somewhat dampened in this case, since the regions of worst case performance by PR are not quite as extreme. NPR consistently remains a very poor performer with the modified measures.

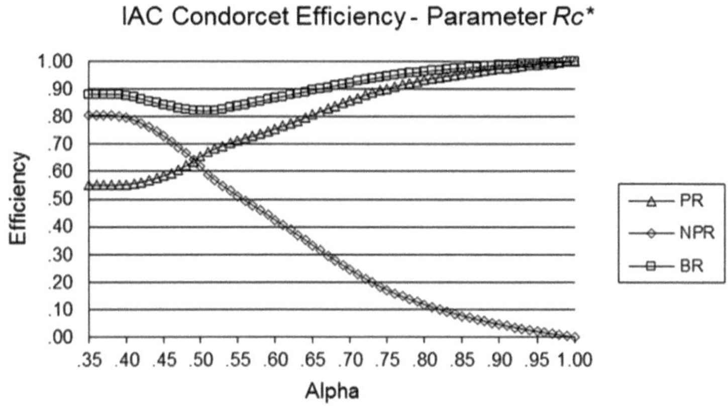

Fig. 4.23 Limiting IAC Condorcet Efficiency of PR, NPR and BR with parameter Rc^*

4.7 Combining Refined Parameters and Modified Databases

The Borda Compromise initially resulted from an analysis that was based on the original definitions of the Measures of Group Mutual Coherence with IAC. When attention was focused on voting situations for which the selection of a voting rule could have an impact on the outcome with MIAC, the differences in the Condorcet Efficiency of PR, NPR and BR were magnified. When Refined Parameters were introduced, there were some dramatic and inconsistent shifts in the patterns of efficiency values for each voting rule that were very different across the Refined Parameters. This had the effect of dampening some of the largest observed differences in efficiencies from earlier studies. Both of these modifications gave good insight regarding important factors that have a significant impact on the evaluation of the relative performances of PR, NPR and BR. The concluding step of this line of analysis was conducted by Gehrlein et al. (2016c) to determine the overall impact that the simultaneous application of both modifications has on the outcome.

4.7.1 Refined Parameters with Modified Bases: Weak Measures

Representations were obtained for the limiting Condorcet Efficiency $CE_{VR}^{WPU}(\infty, MIAC(\alpha_{Rb}))$ for each PR, NPR and BR. This is the worst case scenario for PR efficiency and the results are shown in Fig. 4.24.

The results in Fig. 4.24 show BR dominating PR by a significantly wide margin over the entire range $0 \le \alpha_{Rb} \le \frac{1}{3}$. Moreover, PR efficiencies drop to have values of

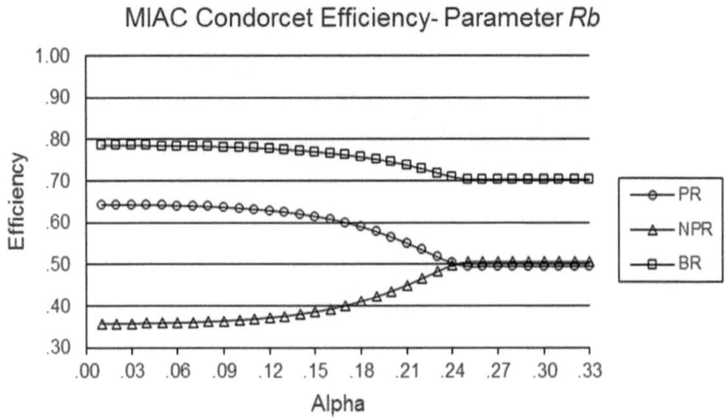

Fig. 4.24 Limiting MIAC Condorcet Efficiency of PR, NPR and BR with parameter *Rb*

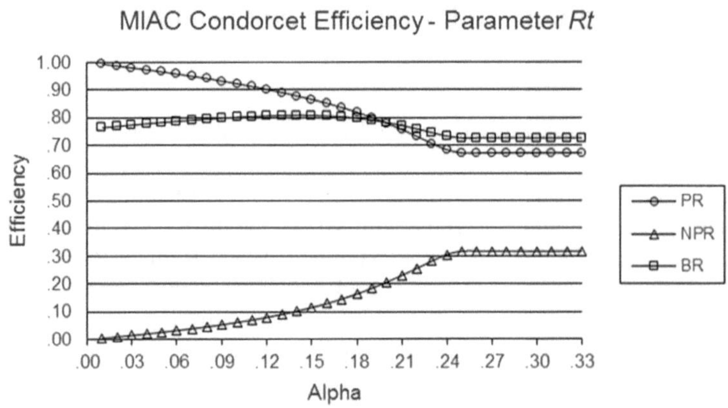

Fig. 4.25 Limiting MIAC Condorcet Efficiency of PR, NPR and BR with parameter *Rt*

only around .50 in the upper range of parameter values. BR also dominates PR in a similar manner for Parameter *Rc*, but the PR efficiencies are improved.

The best PR performance with Refined Weak Measures occurs with Parameter *Rt*, and the results are shown in Fig. 4.25.

The results in Fig. 4.25 show that PR does have greater Condorcet Efficiency than BR in the range $0 \leq \alpha_{Rt} \leq .18$. However, the Condorcet Efficiency of BR never falls below approximately .75, and it has marginally better efficiency than PR in the remaining part of the range. As we have come to expect, NPR performs poorly.

The overall validity of the observations from the best-case performance scenario for PR in Fig. 4.25 was tested against empirically-based simulation estimates for PR efficiency from the Tideman-Plassmann Model. The simulated estimates for PR

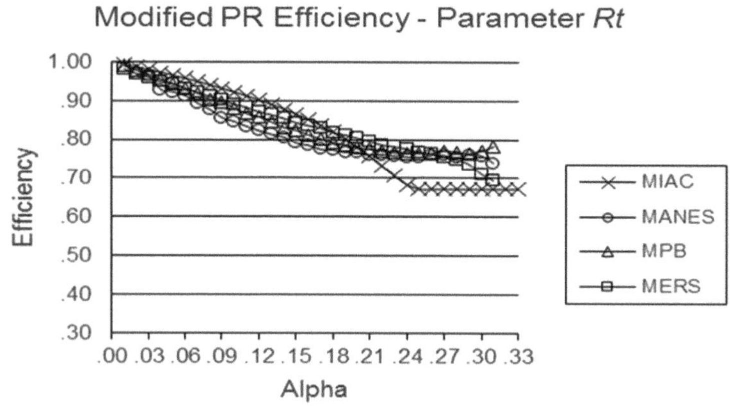

Fig. 4.26 Empirically-based modified Efficiency estimates for PR with parameter *Rt*

efficiency with the modified database for Refined Parameter *Rt* are shown in Fig. 4.26.

The PR efficiencies are reduced for all databases from the MIAC computed results in the region of α_{Rt} where PR dominated BR, and they are improved in the region of α_{Rt} with the lowest MIAC Condorcet Efficiency values for PR.

The results in Fig. 4.27 show the simulated efficiencies for BR, where the efficiency estimates consistently exceed the calculated MIAC efficiencies over the entire range of α_{Rt} for all databases, particularly for the upper end of this range.

So, the impact of having a region where PR dominates BR with MIAC values in Fig. 4.25 is somewhat minimized by the empirically-based results, since they suggest not only that the PR efficiencies are expected to be reduced in that region, but the BR efficiencies should be expected to increase. Moreover, the remaining region where BR already dominated PR is expected to have an increased degree of domination. In the same way, the consistent dominance of BR over PR with Parameters *Rb* and *Rc* is magnified in results from the empirically-based simulations. These overall results reinforce the logic behind the Borda Compromise for Revised Weak Measures with Modified databases.

4.7.2 *Refined Parameters with Modified Bases: Strong Measures*

Representations were obtained for the limiting Condorcet Efficiency of PR, NPR and BR with MIAC for the Refined Strong Measures. We again focus on the comparison of PR to BR and find some very significant results. The best case scenario for PR occurs with Parameter Rt^* and the results are shown in Fig. 4.28.

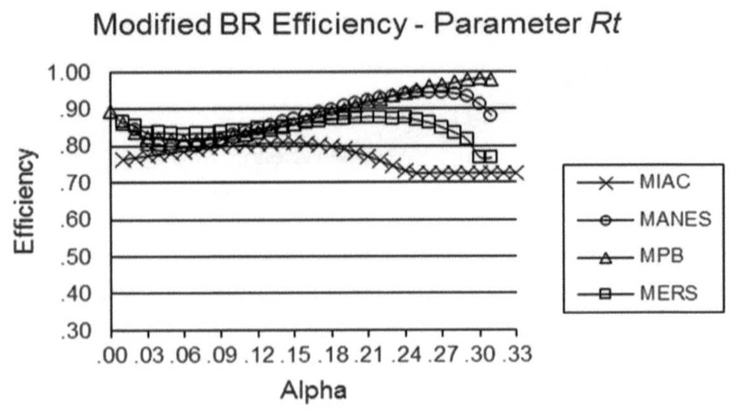

Fig. 4.27 Empirically-based modified Efficiency estimates for BR with parameter *Rt*

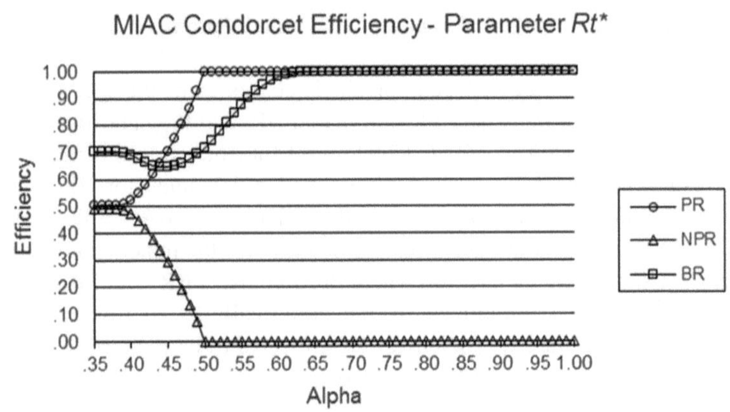

Fig. 4.28 Limiting MIAC Condorcet Efficiency of PR, NPR and BR with parameter Rt^*

The Condorcet Efficiency of PR starts out low at approximately .5 in Fig. 4.28 and then rapidly increases to one, to exceed the BR Efficiency over the range $.36 \leq \alpha_{Rt^*} \leq .60$. However, in this range, the Condorcet Efficiency of BR never falls below approximately .65. NPR has a low level of Condorcet Efficiency over the entire range of α_{Rt^*}, starting at .5 and rapidly decreasing to zero.

The associated PR efficiencies from the empirically-based Tideman-Plassmann Model are shown in Fig. 4.29, where all of the databases show PR efficiencies that are greater than the MIAC estimates in Fig. 4.28. However, PR still starts out with a low level of efficiency at $\alpha_{Rt^*} \to \frac{1}{3}$.

The empirically-based BR efficiencies are shown in Fig. 4.30, where all databases have much greater efficiencies than the MIAC results, with differences that dominate the associated improvements for PR Efficiencies. The patterns of

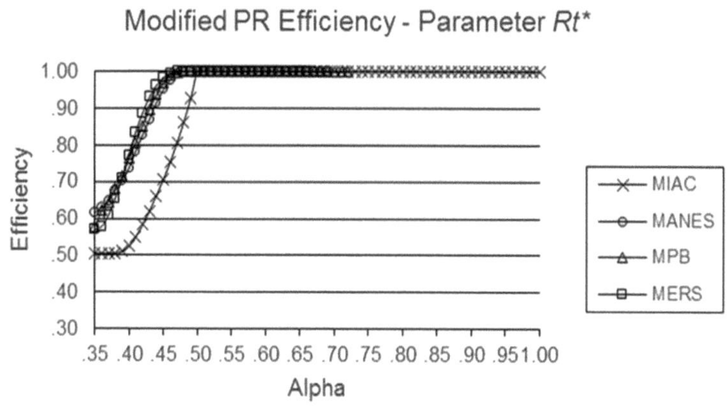

Fig. 4.29 Empirically-based modified Efficiency estimates for PR with parameter Rt^*

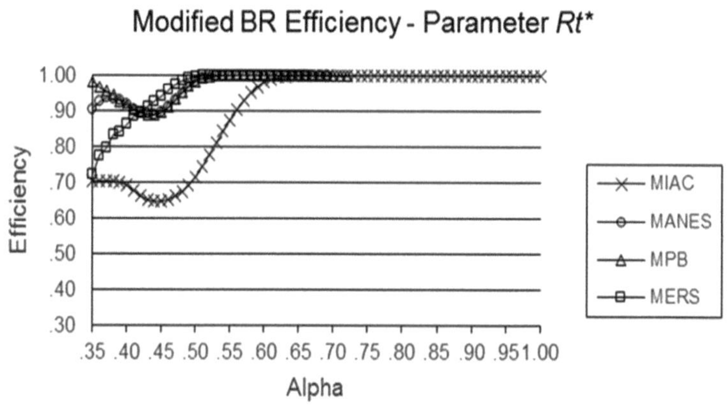

Fig. 4.30 Empirically-based modified Efficiency estimates for BR with parameter Rt^*

efficiency values remain in general agreement with theoretical MIAC results, but the dominance that PR showed over BR for $.36 \leq \alpha_{Rt^*} \leq .60$ is greatly reduced.

We also note here that the empirically-based efficiencies for BR with modified databases look very much the same for Rb^* and Rc^*, which is an important point in the following analysis that ultimately shows that very critical flaws exist with PR. To begin this analysis, the MIAC-based efficiencies for PR, NPR and BR with Parameter Rc^* are shown in Fig. 4.31.

These results show that PR is collapsing over a significant proportion of the range of α_{Rc^*}, with an efficiency of only about .30 for $\alpha_{Rc^*} \to \frac{1}{3}$. Even NPR significantly dominates PR for smaller values of α_{Rc^*}! Validation for this failure of PR is provided with the empirically-based PR efficiencies that are shown in Fig. 4.32.

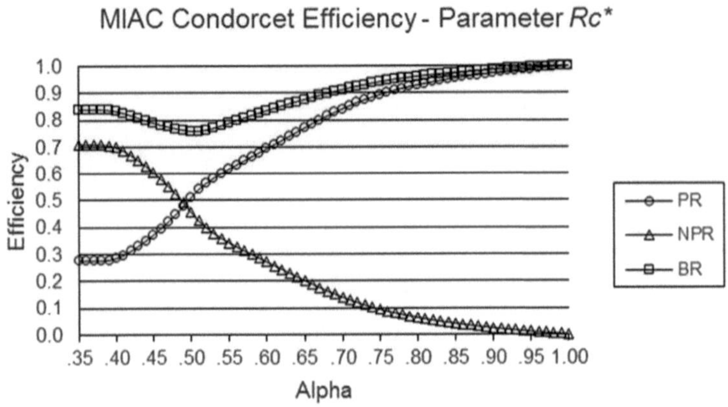

Fig. 4.31 Limiting MIAC Condorcet Efficiency of PR, NPR and BR with parameter Rc^*

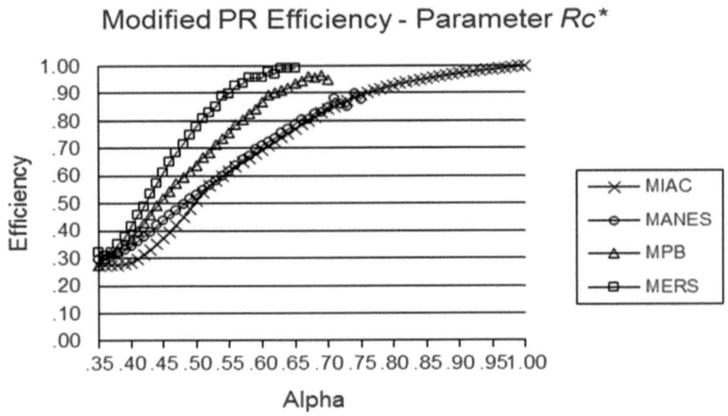

Fig. 4.32 Empirically-based modified Efficiency estimates for PR with parameter Rc^*

The three databases are showing some dispersion in efficiency values in Fig. 4.32, and all estimates are greater than the MIAC-based results. However, the patterns are consistent since they all display very poor performance for PR for smaller values of α_{Rc^*}. And, this is happening while BR is showing very solid performance. So, PR is already looking quite weak compared to BR, and the additional piece of critical information is that PR also shows an almost identical poor performance with Parameter Rb^*. All of the evidence with the combined effects of modified databases and revised parameters yields highly significant support for the Borda Compromise, particularly for the Revised Strong Measures.

References

Black, D. (1958). *The theory of committees and elections*. Cambridge: Cambridge University Press.

Cervone, D., Gehrlein, W. V., & Zwicker, W. (2005). Which scoring rule maximizes Condorcet Efficiency under IAC? *Theory and Decision, 58*, 145–185.

de Condorcet, M. (1788a). On discovering the plurality will in an election. In I. McLean & F. Hewitt (Eds.), *Condorcet: Foundations of social choice and political theory* (pp. 148–156). Hants: Edward Elgar Press.

de Condorcet, M. (1788b). On the constitution and functions of provincial assemblies. In I. McLean & F. Hewitt (Eds.), *Condorcet: Foundations of social choice and political theory* (pp. 157–168). Hants: Edward Elgar Press.

de Condorcet, M. (1788c). On the form of elections. In I. McLean & F. Hewitt (Eds.), *Condorcet: Foundations of social choice and political theory* (pp. 139–147). Hants: Edward Elgar Press.

Diss, M., & Gehrlein, W. V. (2015). The true impact of voting rule selection on Condorcet Efficiency. *Economics Bulletin, 35*, 242–254.

Dodgson, C. L. (1884). *The principles of parliamentary representation*. London: Harrison and Sons Publishers.

Gehrlein, W. V. (2003). Weighted scoring rules that maximize Condorcet Efficiency. In M. R. Sertel & S. Koray (Eds.), *Advances in economic design* (pp. 53–64). Berlin: Springer.

Gehrlein, W. V., & Lepelley, D. (2016). Refining measures of group mutual coherence. *Quality and Quantity, 50*, 1845–1870.

Gehrlein, W. V., Lepelley, D., & Plassmann, F. (2016a). *To rank or not to rank: A summary*. Technical Note posted at: www.researchgate.net/publication/306276950_To_Rank_or_Not_to_Rank_A_Summary

Gehrlein, W. V., Lepelley, D., & Plassmann, F. (2016b). Should voters be required to rank candidates in an election? *Social Choice and Welfare, 46*, 707–747.

Gehrlein, W. V., Lepelley, D., & Plassmann, F. (2016c). Further support for ranking candidates in elections. *Group Decision and Negotiation, 25*, 941–966.

Gehrlein, W. V., Lepelley, D., & Smaoui, H. (2011). The Condorcet Efficiency of voting rules with mutually coherent voter preferences: A Borda Compromise. *Annals of Economics and Statistics, 101*(102), 107–125.

Gehrlein, W. V., & Plassmann, F. (2014). A comparison of theoretical and empirical evaluations of the Borda Compromise. *Social Choice and Welfare, 43*, 747–772.

Lepelley, D., Pierron, P., & Valognes, F. (2000). Scoring rules, Condorcet Efficiency, and social homogeneity. *Theory and Decision, 49*, 175–196.

Moulin, H. (1988). *Axioms of cooperative decision making*. Cambridge: Cambridge University Press.

Saari, D. G. (1990). The Borda dictionary. *Social Choice and Welfare, 7*, 279–317.

Chapter 5
Two-Stage Election Procedures

5.1 Introduction

It was concluded in the preceding chapter that it should generally be expected to be worthwhile to require that voters provide preference rankings for single-stage elections with a small number of candidates, so that BR can be used. This was based on three facts. First, the additional input requirements from voters and the increased implementation cost to determine the winner by BR would not be significant, since it is still a single-stage voting rule like PR and NPR. Second, while the Condorcet Efficiency of BR is not always greater than PR and NPR, it is never significantly dominated; and identifiable scenarios are found to exist for which both PR and NPR exhibit very poor performance. Finally, the likelihood that different winners might result from using different single-stage voting rules was found to be sufficiently high so that the additional effort from using BR results in a real chance of having a positive impact on the outcome of an election. These conclusions were drawn despite the objections that have been raised by some notable researchers to the basic idea of ever requiring voters to rank candidates, since we are only considering cases with a small number of candidates.

The primary purpose of the current chapter is to reconsider those objections, by evaluating the two-stage voting elimination rules PER and NPER, where voters are never required to produce ranked preferences on candidates. The focus of the analysis is to determine how much PER and NPER might dominate BR on the basis of Condorcet Efficiency, to see if the added complexity of using a two-stage procedure might be worthwhile in order to eliminate the ranking requirement. We do note that the use of PER still requires input about preferences on candidates from a subset of voters that is tantamount to obtaining their preference rankings for the case of three-candidate elections. That subset is the group of voters whose most-preferred candidate is eliminated in the first stage. To explain this, let Candidate A be a voter's most preferred candidate, so this voter casts a vote for A in the first stage of PER voting. If Candidate A is then eliminated in the first round of voting,

W.V. Gehrlein, D. Lepelley, *Elections, Voting Rules and Paradoxical Outcomes*, Studies in Choice and Welfare, DOI 10.1007/978-3-319-64659-6_5

this voter must then determine which of B and C is more preferred in the second round. If that candidate is B, then it must be middle ranked in the voter's preferences and C must be the least-preferred candidate. So, that voter has to provide input that would effectively be the same as asking for their preference ranking. The same exact argument applies to voters who have their least preferred candidate eliminated in the first round with NPER. However, this only applies to those subsets of voters, and no voters are ever required to actually produce a complete ranking on the three candidates in any stage of the voting process.

Before we continue this analysis of two-stage voting rules, it is important to point out that this investigation is significantly complicated by that fact that while the use of these rules does indeed eliminate any significant concerns about requiring complete candidate rankings from voters, it unfortunately creates an entirely different set of new concerns. That is, some of the same notable researchers who objected to requiring candidate rankings also presented considerably stronger objections to using any two-stage voting rules. Condorcet was particularly aggressive in dismissing multiple-stage elections (Condorcet 1789, pp. 174–175) when he argued that information from the preliminary vote would be used by voters to form alliances to promote their favored candidates in the successive stages, since the use of such a procedure "simply exposes elections to intrigue and factions."

Condorcet was willing to consider the use of such sequential-elimination procedures only to the extent that they might be used to reduce a set of a large number of preliminary candidates to create a more manageable subset of candidates for further consideration (Condorcet 1788a, p. 154). His specific suggestion was to consider the elimination of all candidates from a single-ballot election who do not receive at least some very small number of votes from the electorate, to allow only the remaining candidates to be considered as being viable. He also supported the notion of having one group of voters in a first stage of voting who would establish a list of viable candidates, while having a totally separate second group of voters determine the winner from that list (Condorcet 1788b, p. 140). However, Condorcet very strongly expressed complete opposition to having the same group of voters being involved in sequential elections with the removal of candidates at each stage. Black (1958, p. 44) presents very similar arguments to discredit the use of any multiple-stage election procedures. Despite these considerable objections, we continue with an evaluation of PER and NPER relative to BR on the basis of the Condorcet Criterion, in order to see what happens.

The same type of analysis that was used to evaluate single-stage voting rules is employed here to determine the expected benefit that might be obtained from increasing the complexity of election procedures to allow the consideration of PER and NPER. Since the conclusion has already been reached that BR should be used as a single-stage rule, we are interested in determining the incremental benefit that can be achieved from using PER and NPER relative to BR. The consistently good performance of BR has already been well established, so we now need to evaluate the relative performance of the two-stage voting rules. The particular point of interest is to determine if there are identifiable scenarios for which each of PER and NPER exhibit poor performance, as we observed previously with PR and NPR.

5.2 Overall Expected Condorcet Efficiencies with IAC

An overall perception of the expected relative performances of BR, PER and NPER on the basis of Condorcet Efficiency can be obtained from looking at their limiting efficiency values as $n \to \infty$ with IAC from Gehrlein and Lepelley (2011), as shown in Table 5.1.

These results strongly suggest that BR will have a much more difficult time in competing against the two-stage elimination rules than it did when it was being compared to PR and NPR. BR had the greatest limiting overall IAC-based Condorcet Efficiency among PR, NPR and BR in Table 4.4, while it now has the lowest corresponding efficiency among BR, PER and NPER in Table 5.1. We already know that BR maintains very stable performance for Condorcet Efficiency as the established measures of Group Mutual Coherence change, and the overall IAC-based results in Table 5.1 do not preclude the possibility that each of PER and NPER could still have identifiable regions of these measures with very poor performance. This possibility must definitely be considered before it is concluded that BR is the obvious loser in its comparison to the two-stage elimination rules.

5.2.1 Condorcet Efficiencies with IAC for Weak Measures

Representations for the limiting Condorcet Efficiency $CE_{VR}^{WPU}(\infty, IAC(\alpha_b))$ as $n \to \infty$ for both PER and NPER are obtained in Gehrlein et al. (2011), and the resulting efficiencies are displayed graphically in Fig. 5.1 along with the associated efficiency values for BR from Fig. 4.2.

The Condorcet Efficiency for PER dominates BR over the entire range $0 \leq \alpha_b \leq \frac{1}{3}$, while NPER dominates BR for $0 \leq \alpha_b \leq .25$ and then declines as α_b increases beyond that range. At the limit as $\alpha_b \to \frac{1}{3}$, the efficiency of NPER is only about .66, so NPER does drop to levels of very poor performance, while BR never allows this to happen. However, this region of failure of NPER does not represent a scenario that is nearly as dramatic as the collapse of NPR in the same region for Parameter b in Fig. 4.2.

Table 5.1 Limiting Condorcet Efficiency values with IAC

Voting rule	IAC efficiency
BR	.9111
PER	.9685
NPER	.9704

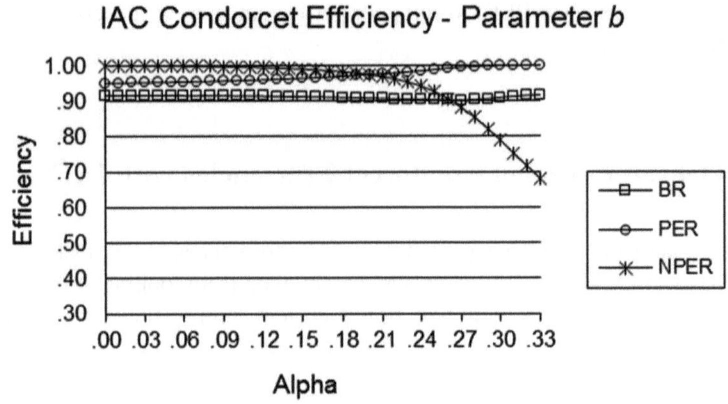

Fig. 5.1 Limiting IAC Condorcet Efficiency of PER, NPER and BR with parameter *b*

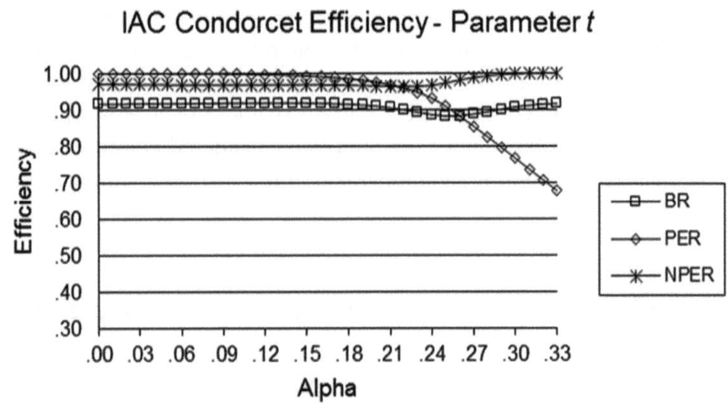

Fig. 5.2 Limiting IAC Condorcet Efficiency of PER, NPER and BR with parameter *t*

The same type of outcome is observed for PER when Parameter *t* is considered, as shown in Fig. 5.2 where the roles of PER and NPER are reversed from the results in Fig. 5.1. NPER dominates BR on the basis of Condorcet Efficiency over the entire range of $0 \leq \alpha_t \leq \frac{1}{3}$, while PER only does so over the range $0 \leq \alpha_t \leq .25$ and then has sharply declining efficiencies as α_t increases beyond that range. As with NPER, this region of failure for PER is not as dramatic as what happens to PR for Parameter *t* in Fig. 4.3.

Condorcet Efficiency results for Parameter *c* are obtained in Gehrlein et al. (2011), but they are not included here, since they are quite similar in nature to the efficiencies for Parameter *b* in Fig. 5.1. The only difference is that the Condorcet Efficiency of NPER declines and falls below BR as $\alpha_c \rightarrow 0$, and the reduction of NPER efficiency is not as strong as it appears in Fig. 5.1.

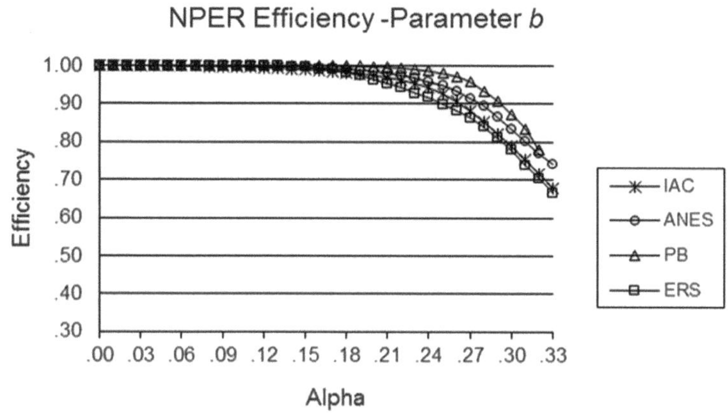

Fig. 5.3 Empirically-based efficiency estimates for NPER with parameter *b*

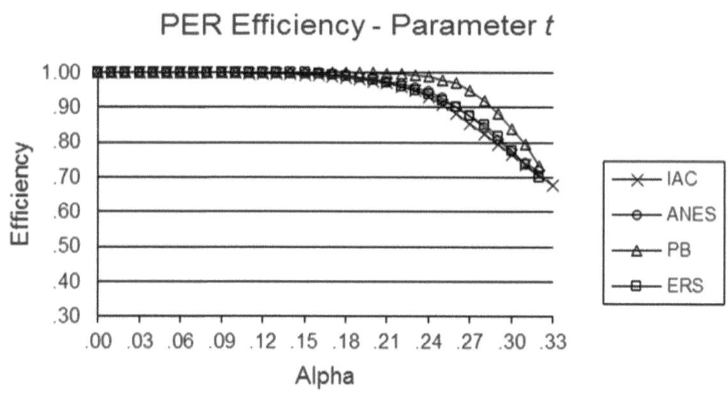

Fig. 5.4 Empirically-based efficiency estimates for PER with parameter *t*

These theoretically-based conclusions were tested with empirically-based simulation analysis by using the Tideman-Plassmann Model in Gehrlein and Plassmann (2014). The Condorcet Efficiency of NPER as related to Parameter *b* are displayed in Fig. 5.3 for the three databases ANES, PB and ERS, along with the corresponding limiting IAC efficiency.

The NPER efficiency estimates are typically greater than the IAC values for all of the databases for Parameter *b* in Fig. 5.3. However, the exact same outcome was established for BR Efficiencies in the preceding chapter. The critical point of interest is that the NPER efficiencies are indeed decreasing sharply to unacceptable levels for all three databases as $\alpha_b \to \frac{1}{3}$.

The same general outcome is observed with the empirically-based simulation estimates for the Condorcet Efficiency of PER with Parameter *t* in Fig. 5.4.

The overall conclusion is that these observations are supportive of the basic principle of the Borda Compromise for Weak Measures. However, this degree of support is clearly dampened considerably when comparing BR to PER and NPER, since the identifiable regions of poor performance for PER and NPER do not result in efficiencies that are as bad as those that are observed with PR and NPR. It will definitely be interesting to see what happens with Strong Measures.

5.2.2 *Condorcet Efficiencies with IAC for Strong Measures*

The extension of this analysis to consider Strong Measures of Group Mutual Coherence starts with representations for the limiting Condorcet Efficiency $CE_{VR}^{SNU}(\infty, IAC(\alpha_{b^*}))$ as $n \to \infty$ for both PER and NPER with Parameter b^* that are obtained in Gehrlein et al. (2011), and the computed efficiencies are displayed in Fig. 5.5 along with the associated efficiency values for BR from Fig. 4.8.

The Condorcet Efficiency results in Fig. 5.5 show that PER dominates BR over the entire range $\frac{1}{3} \leq \alpha_{b^*} \leq 1$ while NPER efficiency dominates BR for $.40 \leq \alpha_{b^*} \leq 1$ and then declines as α_{b^*} decreases below that range. At the limit as $\alpha_{b^*} \to \frac{1}{3}$, the efficiency of NPER is only about .70, so NPER does have an identifiable region of poor performance, while we observed in the preceding chapter that BR never allows this to happen. However, as we just observed with Weak Measures, this region of poor performance for NPER does not represent a scenario that is nearly as dramatic as the collapse of NPR in the same region for Parameter b^* in Fig. 4.8.

The roles of PER and NPER are reversed with Strong Measures for Parameter t^* compared to the results with b^*, just as they were for Weak Measures t and b above. The results for Parameter t^* are displayed in Fig. 5.6 where NPER dominates BR

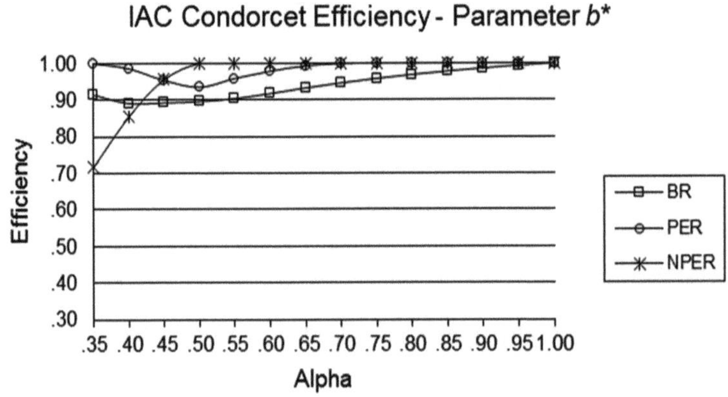

Fig. 5.5 Limiting IAC Condorcet Efficiency of PER, NPER and BR with parameter b^*

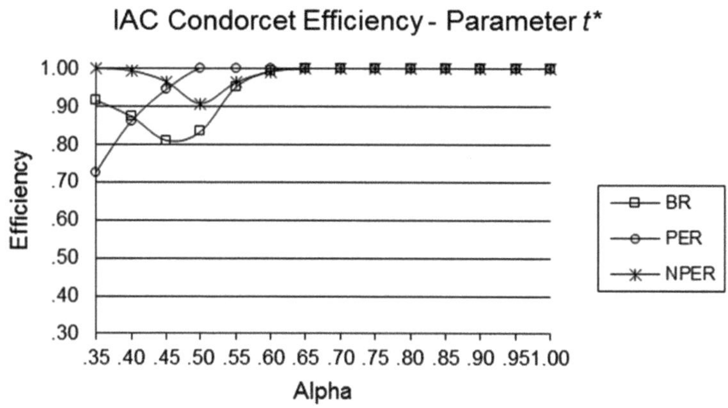

Fig. 5.6 Limiting IAC Condorcet Efficiency of PER, NPER and BR with parameter t^*

over the entire range $\frac{1}{3} \leq \alpha_{t^*} \leq 1$ while PER efficiency dominates BR for $.40 \leq \alpha_{t^*} \leq 1$ and then declines as α_{t^*} decreases below that range. At the limit as $\alpha_{b^*} \rightarrow \frac{1}{3}$, the efficiency of PER is only about .70.

An identifiable region of poor performance clearly exists for PER, but it is very important to note that while BR does have better performance than PER in this region, it also has efficiency values as low as about .80 for α_{t^*} around .45. These theoretically-based results for Parameter t^* in Fig. 5.6 provide an even less enthusiastic endorsement in support of the Borda Compromise than we just observed in Fig. 5.5 for Parameter b^*. However, this disappointing observation for BR Efficiency is somewhat ameliorated by the fact that it has already been determined that the region of low Condorcet Efficiency for BR with Parameter t^*, where its efficiency dips after the starting point at $\alpha_{t^*} = \frac{1}{3}$ and then recovers, largely disappeared with empirically-based simulation analysis from all databases in Fig. 4.12.

Condorcet Efficiency results are also obtained for Parameter c^* in Gehrlein et al. (2011), but they are not included here. However, those results show that the efficiency declines that are observed for NPER and PER in Figs. 5.5 and 5.6 for Parameters b^* and t^* disappear for the case that $\alpha_{c^*} \rightarrow 0$. Given the differences in performance that are observed between theoretical and empirically-based simulation analysis for BR Efficiency, it is definitely of interest to see what happens with empirically-based NPER and PER Efficiencies.

Empirically-based simulation result from Gehrlein and Plassmann (2014) are shown for limiting NPER Efficiency with Parameter b^* and for PER Efficiency with Parameter t^* in Figs. 5.7 and 5.8 respectively, where the results for both parameters are almost identical in nature. The simulated efficiencies are greater than the theoretical results in all cases. However, the low efficiency of about .70 as $\alpha_{b^*} \rightarrow \frac{1}{3}$ for NPER and as $\alpha_{t^*} \rightarrow \frac{1}{3}$ for PER remains consistent in all cases.

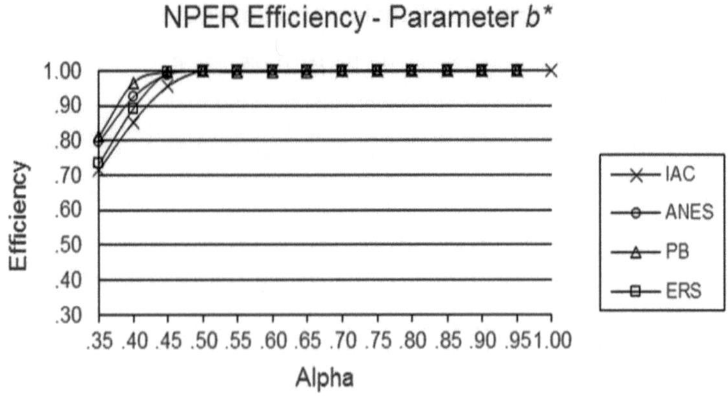

Fig. 5.7 Empirically-based efficiency estimates for NPER with parameter b^*

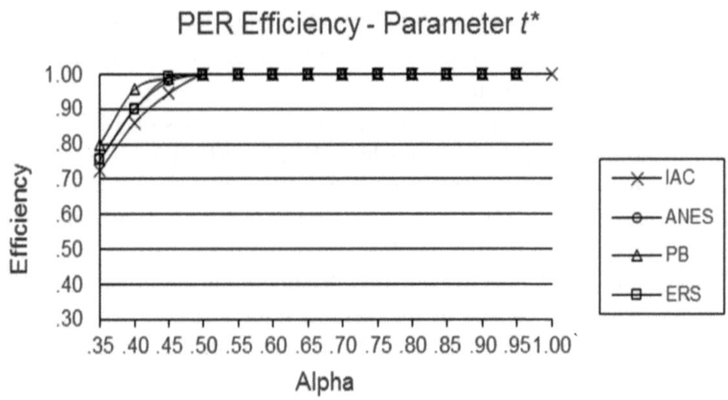

Fig. 5.8 Empirically-based efficiency estimates for PER with parameter t^*

There is support for the Borda Compromise when BR is compared to the two-stage elimination procedures PER and NPER, particularly when evidence from the empirically-based simulations with the Tideman-Plassmann Model is introduced. But, this support definitely is not as strong as what we observed when BR was being compared to the single-stage rules PR and NPR. It is however important to recall that significant changes in Condorcet Efficiency results occurred when both Refined Measures and Modified Bases were introduced during the analysis of single-stage rules. These factors therefore could have a very serious impact on the preliminary conclusions that have been found with the first step of the analysis of two-stage voting rules.

5.3 Efficiencies of Two-Stage Rules with Refined Parameters

When Refined Measures of Group Mutual Coherence were considered for single-stage voting rules, the Condorcet Efficiency differences between PR, NPR and BR were dampened relative to the differences that were observed with the original parameter definitions. The same type of analysis is now used to compare BR to PER and NPER.

The possibility that BR might be able to look better relative to PER and NPER when Refined Measures are considered is made more remote with some initial findings in Gehrlein and Lepelley (2016) that show that PER must elect the CW when t is the $\mathrm{Min}\{b, t, c\}$ and that NPER must do the same when b is the $\mathrm{Min}\{b, t, c\}$, so that $CE_{PER}^{WNU}(\infty, IAC(\alpha_{Rt})) = 1$ and $CE_{NPER}^{WPU}(\infty, IAC(\alpha_{Rb})) = 1$. The Condorcet Efficiency values of BR, PER and NPER for the cases when each of b, t and c is the minimum of the three parameters are summarized in Table 5.2.

The results of Table 5.2 indicate that the possibility that BR might outperform either PER or NPER over some part of the range of associated parameter values is not very promising for either of Parameters Rb and Rt, but there is some chance that this might happen for Parameter Rc.

The same basic outcome results for Refined Parameters of the Strong Measures, where it is found that $CE_{NPER}^{SNU}(\infty, IAC(\alpha_{Rb^*})) = 1$ and $CE_{PER}^{SPU}(\infty, IAC(\alpha_{Rt^*})) = 1$. The Condorcet Efficiency values of BR, PER and NPER for the cases when each of b^*, t^* and c^* is the maximum of the three parameters are summarized in Table 5.3.

The results in Table 5.3 indicate that BR has the worst case scenario for yielding greater Condorcet Efficiency than either of PER or NPER over the range of Parameter Rb^* values. The Efficiencies of all three voting rules are so high with Parameter Rt^* that any differences are likely to be very small as the parameter values are changed. So, the best case scenario for BR to have greater Condorcet Efficiency than PER or NPER occurs within the range of Parameter Rc^* values.

5.3.1 *Two-Stage Efficiencies with Refined Weak Measures*

Representations were obtained for the limiting Condorcet Efficiency of PER and NPER with the IAC assumption for specified values of Refined Weak Measures in Gehrlein and Lepelley (2016). Computed efficiencies for PER and NPER from

Table 5.2 Limiting IAC Efficiencies with refined parameters for weak measures

	Minimum parameter		
Voting rule	b	t	c
BR	.9065	.9065	.9211
PER	.9539	1	.9503
NPER	1	.9783	.9298

Table 5.3 Limiting IAC efficiencies with refined parameters for strong measures

Voting rule	Maximum parameter b^*	t^*	c^*
BR	.8795	.9679	.8829
PER	.9920	1	.9069
NPER	1	.9786	.9279

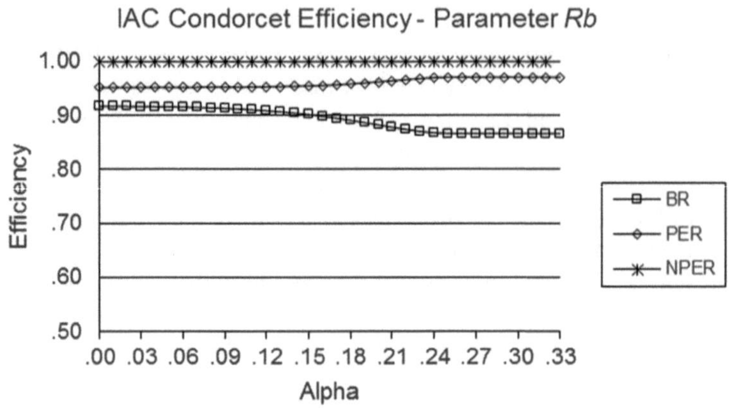

Fig. 5.9 Limiting IAC Condorcet Efficiency of PER, NPER and BR with parameter *Rb*

these representations are displayed graphically along with the associated BR Efficiencies in Fig. 5.9 for one of the worst case scenarios for BR, as described above, with Parameter *Rb*.

The results in Fig. 5.9 are consistent with the expectations from Table 5.2, with NPER dominating PER, which in turn dominates BR. The phenomenon where the Condorcet Efficiency of NPER plunges for larger values of Parameter b in Fig. 5.1 vanishes with the use of Refined Parameter *Rb*, to erase the benefit of using BR in that case. The results for Parameter *Rt* were found to be very similar, with the roles of PER and NPER being reversed, such that the drop in PER Efficiency for large values of Parameter t in Fig. 5.2 disappears.

If BR has any chance of performing with greater Condorcet Efficiency than either of PER or NPER with any Refined Weak Measure, it must therefore occur with Parameter *Rc*, and the efficiencies of the three voting rules for that scenario are shown in Fig. 5.10. In this case, PER beats BR over the entire range $0 \leq \alpha_{Rc} \leq \frac{1}{3}$. However, BR does have marginally better performance than NPER over parameter values in the range $0 \leq \alpha_{Rc} \leq .10$, while NPER efficiencies increase and BR Efficiencies decrease as α_{Rc} increases above that range. But, this is only one small victory for BR, after some significant setbacks, so it becomes critical to see what happens with the Refined Strong Measures.

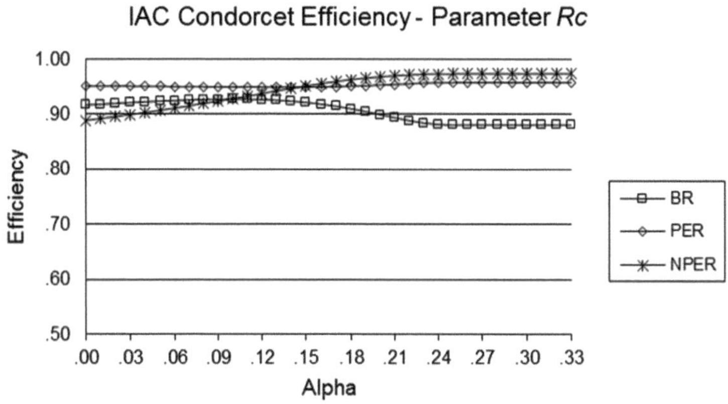

Fig. 5.10 Limiting IAC Condorcet Efficiency of PER, NPER and BR with parameter Rc

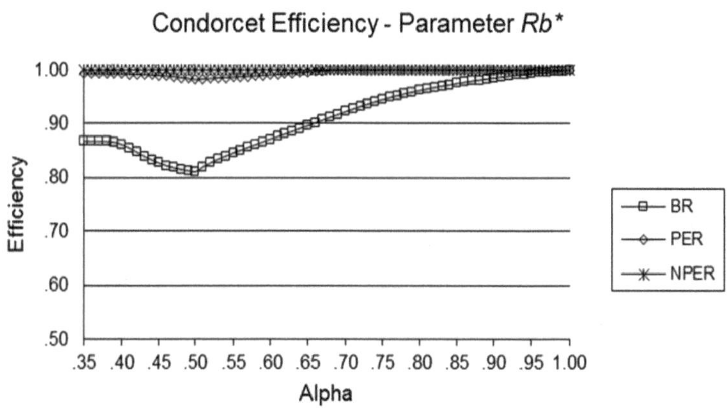

Fig. 5.11 Limiting IAC Condorcet Efficiency of PER, NPER and BR with parameter Rb^*

5.3.2 *Two-Stage Efficiencies with Refined Strong Measures*

The evidence that is presented in Table 5.3 suggests that the worst case scenario for BR in its comparison to PER and NPER for Refined Strong Measures exists for Parameter Rb^*, and the best case exists for Rc^*. The Condorcet Efficiencies for BR, PER and NPER are displayed graphically in Fig. 5.11 for Parameter Rb^*.

The results in Fig. 5.11 predictably show perfect performance for NPER at selecting the CW and near perfect performance with PER over the entire range $\frac{1}{3} \leq \alpha_{Rb^*} \leq 1$, so that both PER and NPER significantly dominate BR on the basis of Condorcet Efficiency over the entire range of α_{Rb^*}. This is not at all surprising, given the results in Table 5.3. The significantly declining performance of NPER in Fig. 5.5 for Parameter b^* as $\alpha_{b^*} \to \frac{1}{3}$ from above, that allowed BR to dominate NPER in that range, vanishes for the case of Rb^* in Fig. 5.11. The same result was found to occur for PER with the transfer from Parameter t^* to Rt^* in Gehrlein and Lepelley (2016). This overall outcome is identical to what we observed earlier when BR lost this same type of advantage that it had over PR and NPR in the transition from Weak Measures to Refined Weak Measures.

The Condorcet Efficiency results are shown in Fig. 5.12 for the best case scenario for BR with Parameter Rc^*, where BR is dominated by NPER over the entire range $\frac{1}{3} \leq \alpha_{Rc^*} \leq 1$. But, the degree of dominance of NPER is reduced from what was observed for Parameter Rb^* in Fig. 5.11. It also turns out that BR does indeed have marginally greater Condorcet Efficiency than PER for $.62 \leq \alpha_{Rc^*} \leq 1$. However, PER dominates BR by a greater degree for Parameter Rc^* values below that range.

The comparison of BR to the single-stage rules PR and NPR on the basis of Condorcet Efficiency led to similar conclusions when Refined Parameters replaced the original measures of Group Mutual Coherence, but the differences in the relative comparisons of efficiencies became somewhat reduced as a result of that transition. The same type of comparison of BR to the two-stage rules PER and NPER is quite different, and significant changes result with the transition to Refined Parameters. BR is typically found to be dominated by both PER and NPER for both Refined Weak and Strong Measures, with only a few exceptions. BR still generally performs well in all situations, but PER and NPER are almost always performing even better. Moreover, the existence of identifiable regions of very poor performance with PER and NPER vanishes with the transition to Refined Measures.

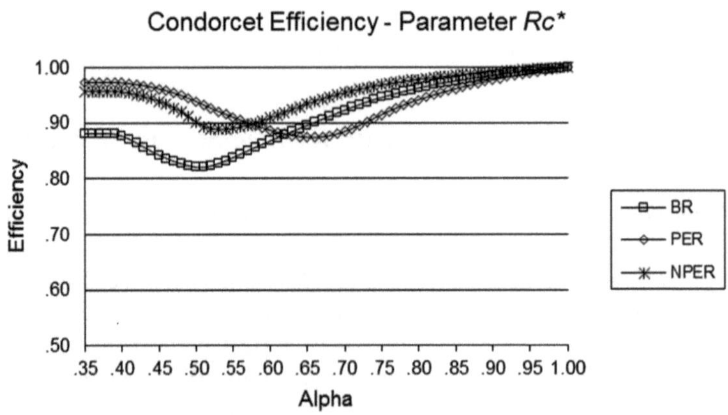

Fig. 5.12 Limiting IAC Condorcet Efficiency of PER, NPER and BR with parameter Rc^*

5.4 Two-Stage Rules: Refined Parameters and Modified Bases

During the earlier analysis of single-stage voting rules in the preceding chapter, a Modified Base was created that eliminated all voting situations in which all $WSR(\lambda)$ elect the same winner, in order to focus on the remaining voting situations for which the selection of a voting rule might have some actual impact on the outcome of an election. This was achieved by considering only the set of voting situations for which PR and NPR do not elect the same winner. The overall effect from doing this was to see Condorcet Efficiency results that magnified the differences between PR, NPR and BR to strongly reinforce the reasoning that led to the Borda Compromise. So, by bringing this option into the current analysis of comparing BR to the two-stage elimination rules PER and NPER, it could happen that BR might recover from the setbacks that just occurred while considering the impact of using Refined Parameters.

A direct application of that same procedure to obtain a Modified Base to compare BR to PER and NPER unfortunately becomes very problematic in this case. The primary issue from doing this arises as a result of voting situations like the one shown in Fig. 5.13 with $n = 11$ voters:

It is easy to verify that both PR and NPR elect Candidate B as the winner for this example, so it is a voting situation that would be removed while creating the Modified Base. However, Candidate A is the winner with both PER and NPER, because both eliminate C in the first round, and then AMB by 6-5 to be the winner in the second round for both of these rules. Candidate A is also the CW with AMB (6-5) and BMC (6-5). So, by eliminating this voting situation in a comparison of BR to PER and NPER, we would be removing a voting situation in which PER and NPER both elect the CW, while BR does not do so, since BR must elect B as a result of the fact that it is the winner by both PR and NPR. This outcome creates a bias in favor of BR, and it is in sharp contrast to our earlier approach, where the elimination of voting situations in which PR and NPR choose the same winner ensured that BR would also choose that same winner.

A solution to this problem is presented in Gehrlein et al. (2017), where the comparison of BR to PER and NPER takes place in two steps. In the first step, all voting situations for which BR and PER elect the same winner are removed for the comparison of those two voting rules. The IAC-based analysis which assumes that all voting situations in the remaining Modified Base are equally likely to be observed is denoted by $MIAC_{BP}$. In the second step, all voting situations in which BR and NPER elect the same winner are removed for the comparison of those two rules, with the associated IAC-based model denoted as $MIAC_{BN}$. The overall

Fig. 5.13 An example voting situation with 11 voters

A	B	C	B
B	A	A	C
C	C	B	A
$n_1 = 4$	$n_3 = 2$	$n_4 = 2$	$n_5 = 3$

Table 5.4 Condorcet Efficiencies with $MIAC_{BP}$ and $MIAC_{BN}$

Voting rule	IAC-based assumption	
	$MIAC_{BP}$	$MIAC_{BN}$
BR	.2428	.1927
PER	.7572	–
NPER	–	.8073

Condorcet Efficiency values that are obtained for BR, PER and NPER with these two models are summarized in Table 5.4.

The terribly low Condorcet Efficiency results with BR for both cases in Table 5.4 very strongly suggest that a dismal performance can be expected for BR when it is compared to either PER or NPER over the ranges of parameter values for measures of Group Mutual Coherence, given the earlier results that followed from Table 5.1 where basic IAC analysis was being used. Given that expectation, we proceed to look at theoretically based limiting Condorcet Efficiency results with $MIAC_{BP}$ and $MIAC_{BN}$ to see what happens as Refined Parameter values change for the Measures of Group Mutual Coherence.

5.4.1 Comparing BR to PER for Refined Parameters with MIAC

The comparison of BR to PER in this scenario is simplified by a result from Gehrlein et al. (2017) that proves that if all voting situations are removed when both of these voting rules elect the same winner, then for limiting Condorcet Efficiency values with any Refined Parameter Rx:

$$CE^{BR}_{MIAC_{BP}}(\alpha_{Rx}) + CE^{PER}_{MIAC_{BP}}(\alpha_{Rx}) = 1.$$

Moreover, this general relationship is valid for all databases, not just for $MIAC_{BP}$, so that BR is dominated by PER whenever the BR Efficiency is below .50. One cause of the very low Condorcet Efficiency of BR in Table 5.4 stems from the fact that its efficiency is found to be equal to zero for all values of the Refined Parameter Rt. Representations were obtained for the limiting Condorcet Efficiency of BR for the two other Refined Weak Measures Rb and Rc, and the resulting computed values are displayed in Fig. 5.14.

These values of the Condorcet Efficiency of BR in Fig. 5.14 are less than .50 for the entire range of all three Refined Weak Measures, so it is completely dominated by PER over every scenario.

When attention is shifted to Refined Strong Measures, things improve very slightly for BR. The BR Efficiency is still shown to be equal to zero over the entire range of Parameter Rt^*, and the Condorcet Efficiency of BR for Refined Parameters Rb^* and Rc^* are displayed in Fig. 5.15.

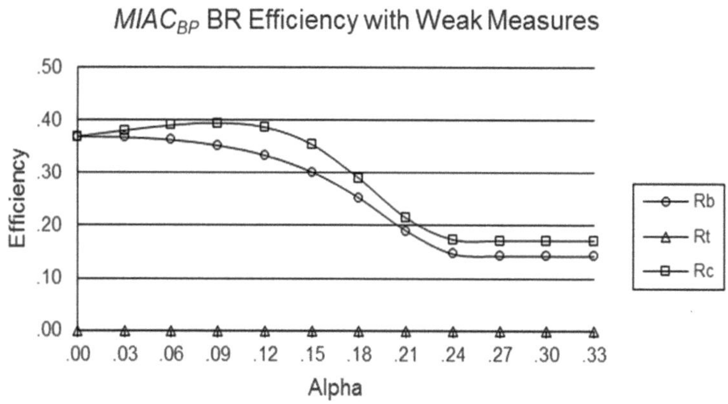

Fig. 5.14 BR efficiency for weak measures when BR and PER choose different winners

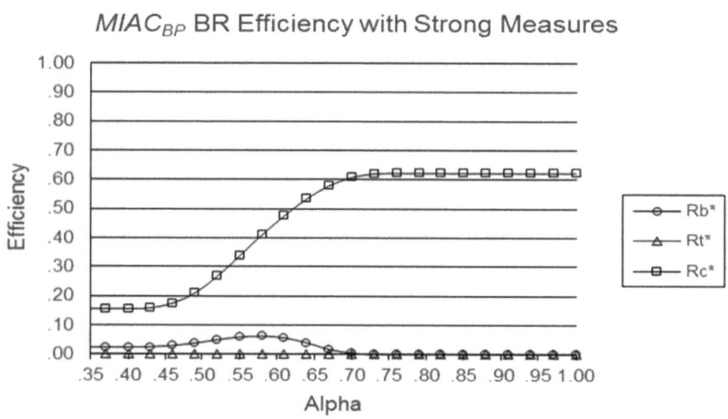

Fig. 5.15 BR efficiency for strong measures when BR and PER choose different winners

The very limited good news for BR with Refined Strong Measures in Fig. 5.15 is that it marginally beats PER over the range $.60 \leq \alpha_{Rc^*} \leq 1$. So, PER completely dominates BR on the basis of Condorcet Efficiency when using theoretically-based analysis with $MIAC_{BP}$ for all three Refined Weak Measures, and BR only marginally beats PER over part of the range of one of the three Refined Strong Measures. This is all very bad news for BR, so we have to wonder if it is finally time to fully accept this evidence to conclude that PER clearly is a much better option than BR, and to reject BR since it really is worth the extra effort to use PER instead.

Before actually taking this final very conclusive step, we momentarily digress and recall some historical events that involved the American humorist Mark Twain, who was touring in London in May of 1897 when reports were widely circulated that he was very ill and was near death. A journalist sent two telegrams to verify this story, to which Twain responded "The report of my death is exaggerated." In a similar episode, it was reported in a newspaper article in May of 1907 that he had been lost at sea, to which he replied in a letter "I will make an exhaustive investigation of this report." Since Mark Twain was able to personally respond to two alarming reports about his death, this gives a glimmer of hope that further investigation could possibly provide some insight to allow us to conclude that "The report of the demise of BR is exaggerated."

Our renewed analysis begins by reiterating what happened when the Borda Compromise was evaluated for single-stage voting rules by comparing BR to PR and NPR. The very critical point was made that this concept was particularly relevant because there was a significant likelihood that the determination of which of these voting rules is to be used could make an actual difference in the election outcome. In particular, it was found that the probability that PR and NPR, and therefore all $WSR(\lambda)$, elect the same winner was .5082 for 101 voters with IAC in Table 4.3; so there was a probability of .4918 that the voting rule selection could make a difference. The subsequent removal of these most concordant voting situations for which PR and NPR elect the same winner resulted in the Modified Base MIAC that was composed of voting situations with lower levels of concordance. This had a significant impact on the resulting Condorcet Efficiency for voting rules with MIAC, compared to the original results with IAC.

The probability that PR and NPR both elect the same winner was then found to be even greater in the empirically-based simulation results, so that greater proportions of the most concordant voting situations were being removed to form their Modified Bases. This outcome promoted scenarios to allow the particular structural differences that are embedded within those databases to have even more of an impact on the voting situations that remained in the Modified Bases. This led to varying degrees of magnification of the Condorcet Efficiency differences that were obtained when comparing the original and modified databases. What happens if we extend this same type of analysis to the comparison of BR and PER?

Gehrlein et al. (2017) finds that the limiting probability as $n \to \infty$ that BR and PER elect the same winner with IAC is given by .8954, so the likelihood that the process of selecting which voting rule to use might have an impact on the election outcome is now greatly reduced to only .1046. Moreover, if the empirical databases lead to greater probabilities that the same winner is elected by both BR and PER, the very large proportions of voting situations that are being removed to modify the databases will result in Modified Bases that are exposed to a dramatically increased impact from the various structural differences that are present within the different databases.

5.4.2 Empirical Comparison of BR to PER for Refined Parameters

Our analysis of what happens when BR is compared to PER with Refined Parameters begins with results from Gehrlein et al. (2017) for Weak Measures that are

Table 5.5 Distribution of parameters for weak measures with BR-PER comparison

	$MBase_{BP}$			
Min$\{b, t, c\}$	IAC	ANES	PB	ERS
b	40.87%	39.81%	35.30%	29.60%
t	28.57%	15.67%	42.68%	57.86%
c	30.56%	44.52%	22.02%	12.54%
Same winner	89.54%	93.24%	96.83%	97.19%

given in Table 5.5, where some very striking observations can be made. This table presents empirically-based simulation results from the Tideman-Plassmann Model with the three databases, and the primarily important piece of information that it includes is the percentage of times that BR and PER both elect the same winner.

The results show that the choice between using BR or PER only makes a difference in the election outcome about 2.8% of the time with ERS, and the maximum impact among the empirical databases occurs with ANES at 6.8%. With such a low probability that using PER will make a difference, compared to using BR, it is now starting to look much easier to suggest that it really could be unlikely to be worth the additional effort and increased resources that are required to implement PER, relative to simply using BR. The tide has started to turn to be a bit more in favor of BR!

A second critical observation from Table 5.5 is that the distributions that show when each of the Parameters b, t and c is the minimum of the three are dramatically different across the different Modified Bases, which we denote by $MBase_{BP}$. It is very evident that since we now eliminate at least 90% of the most concordant voting situations during this transition to each $MBase_{BP}$, the particular intricacies of the models that govern the formation of voters' preferences on candidates come to the forefront to have a very significant impact on the nature of the preferences of voters in the remaining voting situations. Less consistency in Condorcet Efficiency values should therefore be expected across the different modified databases than what was observed across the original databases.

The Tideman-Plassmann Model was used to estimate the Condorcet Efficiency of BR for the Refined Parameters with each $MBase_{BP}$, and the BR Efficiency remains at zero for Parameter Rt. The impact on BR Efficiency that results from the significant differences between the distributions of the Weak Measures across the $MBase_{BP}$ in Table 5.5 is evident for Parameter Rb in Fig. 5.16.

The plotted BR Efficiencies show almost no consistency across the Modified Bases in Fig. 5.16. However, BR is now found to beat PER over parts of the ranges for both $MANES_{BP}$ and MPB_{BP}, as opposed to its previously observed complete dominance by PER with $MIAC_{BP}$

The differences in these shifts of Condorcet Efficiency values across the different $MBase_{BP}$ become even more pronounced when Parameter Rc is considered in Fig. 5.17.

The remarkably inconsistent patterns of Condorcet Efficiency in Fig. 5.17 shows BR significantly dominating PER over the entire range of $MANES_{BP}$ and over most of the range of MPB_{BP}, while PER dominates BR over the entire range of $MERS_{BP}$.

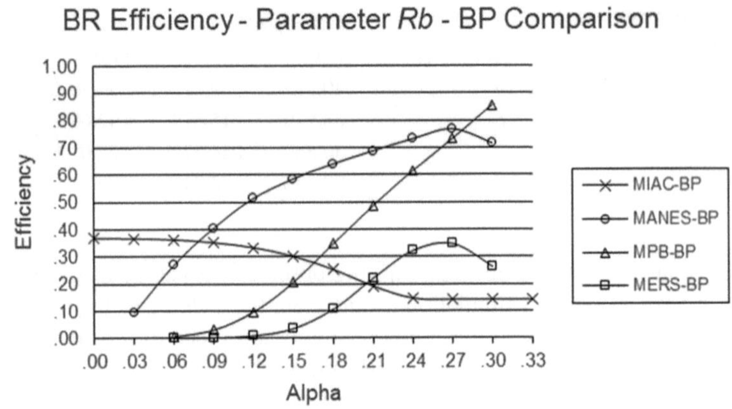

Fig. 5.16 BR efficiency estimates for parameter *Rb* when BR and PER elect different winners

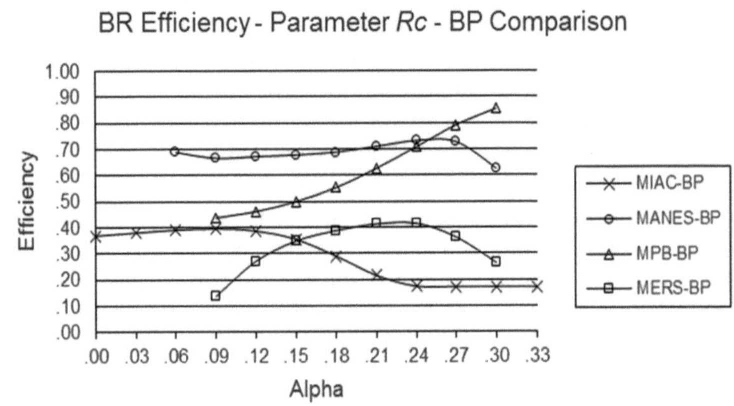

Fig. 5.17 BR efficiency estimates for parameter *Rc* when BR and PER elect different winners

The theoretical $MIAC_{BP}$ results that made BR look so bad in its comparison to PER for the Refined Weak Measures in Fig. 5.14 are not really wrong, but they can be misleading. Because the particular intricacies of the models that are governing the formation of voters' preferences on candidates in each database are magnified while creating the various $MBase_{BP}$, there are very significant differences in the basic internal structure for the preferences of voters in the voting situations that are retained in each case. None of the databases produces BR Efficiency results that look anything like the results for any other database, and the Condorcet Efficiency of BR does in fact look very good relative to PER in some cases.

These empirically-based simulation results for Refined Weak Measures do indeed tend to make BR look somewhat more appealing than the earlier theoretical $MIAC_{BP}$ analysis suggested, but BR still generally tends to be dominated by PER. This observation is particularly enhanced by the fact that the BR Efficiency is zero

Table 5.6 Distribution of parameters for strong measures with BR-PER comparison

	$MBase_{BP}$			
Max$\{b^*, t^*, c^*\}$	IAC	ANES	PB	ERS
b^*	37.86%	24.35%	57.43%	78.08%
t^*	9.94%	5.88%	7.11%	5.05%
c^*	52.19%	69.76%	35.46%	16.87%

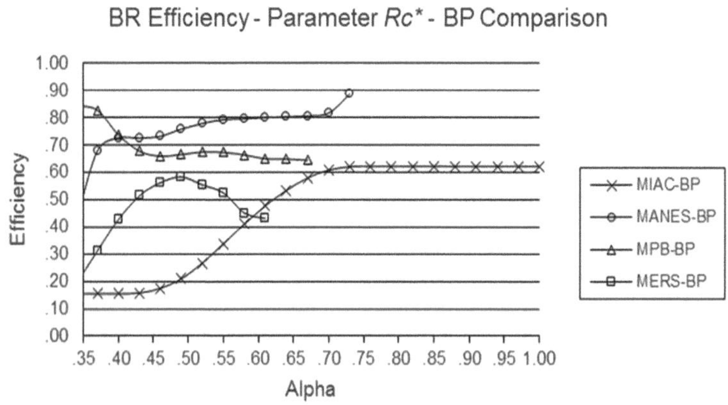

Fig. 5.18 Simulated BR efficiency estimates for parameter Rc^* in BR-PER comparison

for Parameter Rt, and the probability that t is the Min$\{b, t, c\}$ can be significant in Table 5.5. However, this general dominance must be considered in light of the very important fact that there is a relatively small chance that PER will actually choose a different winner than the one that is selected by BR.

The extension of this analysis to consider Refined Strong Measures begins with the results in Table 5.6 that show the distributions of those parameters for the $MBase_{BP}$.

The BR Efficiency is equal to zero over the entire range of Parameter Rt^*, but the impact of that fact is significantly minimized for the Revised Strong Measures, since the results in Table 5.6 consistently show that t^* is the Max$\{b^*, t^*, c^*\}$ for only a relatively small percentage of cases. The Condorcet Efficiency values for BR are also found to be dominated by PER with Parameter Rb^* in almost every case, which is not a very surprising result, given the very small associated BR Efficiencies in Fig. 5.15.

The BR Efficiencies for Refined Parameter Rc^* are displayed in Fig. 5.18, where BR is found to perform very well. It dominates PER over the entire range of $MANES_{BP}$ and MPB_{BP}. It also dominates PER over parts of the ranges for both $MERS_{BP}$ and $MIAC_{BP}$.

The overall Condorcet Efficiency results for the Refined Strong Measures generally tend to make BR look a bit more appealing than the results that were observed for Refined Weak Measures, since BR actually outperforms PER for some ranges of the refined parameters for both the theoretical and empirically-based

Table 5.7 The likelihood that PER and NPER elect the same winner

Database	Same winner with PER-NPER
IAC	94.27%
ANES	93.24%
PB	99.36%
ERS	98.28%

simulation results for the Refined Strong Measures. We also note again for emphasis that there is only a small chance that PER will actually choose a different winner than BR to begin with.

5.4.3 *Other Comparisons*

Other voting rule comparisons were performed in Gehrlein et al. (2017), with the primary effort being an evaluation of the Condorcet Efficiency of BR relative to NPER. The results of that comparison are not included here, since the nature of those results very closely mirrors what we have just observed when comparing BR to PER.

A comparison of PER and NPER was also performed to account for the possibility that there might be some scenarios in which the objection to requiring candidate rankings from voters might be so strong that these are the only two options to be considered. Following the logic of earlier discussion, the $MIAC_{PN}$ assumption was developed to account for the case in which all voting situations for which PER and NPER do not elect the same winner are equally likely to be observed. The limiting Condorcet Efficiency with $MIAC_{PN}$ was found to be .4849 for PER and .5151 for NPER, so that no overall domination of one of these two rules over the other can be expected on the basis of Condorcet Efficiency.

The percentage of times that PER and NPER both elect the same winner was also obtained for each database, and those results are summarized in Table 5.7. These results clearly indicate that the likelihood is remarkably small that PER and NPER will produce different winners.

5.5 Other Relevant Factors and Conclusion

When attention is restricted to voting situations in which the voting rules do not elect the same winner, the two-stage elimination voting rules PER and NPER typically, but not always, have a higher Condorcet Efficiency than BR. But, the very critical factor is that the likelihood that either PER or NPER will choose a different winner than the one that is selected by BR is quite likely to be too small to make the additional effort of using them worthwhile. This particular observation is very strongly reinforced in the empirically-based results.

It has been routinely mentioned that the use of the two-stage election rules requires additional effort and increased resources, but no actual costs have been provided. It would be very difficult to attach some precise monetary value to what resources are required to obtain the associated increase in Condorcet Efficiency that results from using a two-stage procedure, compared to using a single-stage election procedure. It is however very clear that this cost can be very substantial. Gehrlein et al. (2017) provide some documentation of this regarding the cost of the 2012 two-stage French Presidential election, which was 228 million EUR. When the various elements of the costs that are associated with that particular election are considered, the additional cost that was incurred from using a two-stage procedure rather than a one-stage procedure falls roughly between 50 million EUR and 70 million EUR.

It is worthwhile to consider an evaluation of BR and the two-stage elimination rules on the basis of their susceptibility to exhibiting other possible voting paradoxes. It has already been noted that BR cannot elect the CL, to thereby avoid any possible occurrence of a Weak Borda Paradox, as discussed in first chapter. However, since the second stage of both PER and NPER voting is based on Majority Rule on the remaining two candidates, this leads to the obvious conclusion that neither one of these two elimination rules can exhibit an example of a Weak Borda Paradox either.

A particular disadvantage that arises from using the two-stage elimination procedures is that they become vulnerable to exhibiting various Monotonicity Paradoxes, as evidenced by the example voting situation that is provided in Fig. 1.18 that used NPER to illustrate the Monotonicity-based No Show Paradox. However, BR is fortunately immune to displaying any Monotonicity Paradoxes. This is a significant issue, since Miller (2016) concludes from simulation analysis that failures of monotonicity definitely should not be viewed as rare events in closely contested two-stage elections with three candidates. Lepelley et al. (2015) reach the same general conclusion from a theoretical analysis that was based on IAC.

A number of studies have been conducted to find that BR uniquely exhibits some very positive properties. See for example Young (1974) and Fishburn and Gehrlein (1976). Another series of impressive studies by Donald Saari have also shown that BR is very effective at minimizing the probability that a number of voting paradoxes will be observed (Saari 1989, 1990, 1996, 1999). Most of these analyses are very technical in nature, and it suffices to say here that there are many other good reasons to like BR. But, it is very important to note that there is a major criticism of BR in the literature that is related to its susceptibility to manipulation, where voters can change the outcome of an election by misrepresenting their true preferences. We are not ignoring this problem, but instead will present a full consideration of this criticism in the final chapter.

It definitely might be argued that the additional cost and effort that is required to use a two-stage election procedure could be justified for very important elections like the choice of a national leader. Similarly, in elections that start out with large numbers of candidates, it might be worthwhile to use some elimination process to reduce the total set of candidates down to two or three frontrunners. However, an interesting confounding issue to consider here is that Wright and Riker (1989) did an empirical analysis of two-stage election procedures that are used for some elections in the US and found that two additional candidates typically can be expected to enter into an election, just because such a two-stage rule is being used. These candidates would not have entered if PR was being used in a single-stage election, where they perceived no chance of winning.

It was noted in the introduction to this chapter that even Condorcet was supportive of some very limited forms of this approach. However, once the field of candidates has been reduced to three, there is no precise answer regarding whether or not it is worthwhile to proceed with using two-stage elimination procedures, rather than simply to rely on BR. The major factor is that the probability that the winner that is elected by PER or NPER will be different than the candidate that is chosen by BR is so low in this case that the expected benefit that would be obtained from using a two-stage procedure on the reduced set of candidates is small, and very probably is not worth the extra cost and effort.

We conclude our discussion of two-stage voting procedures by considering options like Instant Runoff Voting (IRV) that attempts to obtain the benefits of a two-stage voting rule, while eliminating the need to hold two separate elections. IRV requires voters to rank the candidates on the ballot, and these rankings are then used by the ballot processors to determine the winner by PER from the reported rankings. Due to the way that IRV works, it is subject both to the objection that voters should not be required to rank candidates and to some of the potential objections that two-stage procedures should not be used, despite the fact that voters have to cast only one ballot. A variation of this procedure is used for example in Presidential elections in Ireland. A direct application of IRV that is called "Ranked-Choice-Voting" was adopted in 2016 by a referendum that was held in the State of Maine (US). The proposal passed by a slim margin with 52% of votes approving its adoption. The movement to adopt this voting rule gained significant momentum when an unpopular candidate Paul LePage was elected as Governor of Maine in 2008, and was subsequently re-elected in 2012. In both cases, there were multiple candidates and the winner was determined directly by simple PR.

Strong opposition existed to the adoption of Ranked-Choice-Voting, and Seelye (2016) lists arguments that were used to oppose it. First, this more complicated voting process might depress voter turnout and cause voting errors on the ballots. Second, because of the rural nature of Maine, half of the voting districts still count ballots by hand to further increase the possibility of errors during tabulation. Third, Ranked-Choice Voting could elect a candidate "not favored by the majority," which sounds like concerns were being raised that a Strong Borda Paradox might still occur, as it probably did with PR in the LePage elections. Finally, it is noted that the election procedure can be more costly to implement.

A rather sinister view of the underlying motivation behind the referendum effort to adopt Ranked-Choice-Voting in Maine is presented by Santucci (2016), who notes that it might seem unusual that this reform was promoted by the political establishment, when it might seem more likely to have been initiated by political activists. However, it is asserted that this pattern is consistent with similar movements, since 24 US cities adopted voting rules like Ranked-Choice-Voting during the twentieth century, and that this was accomplished by election reformers who joined forces with a political party that was attempting to regain political power in at least 17 of those cases. After all of this, the Maine Supreme Court declared Ranked Choice Voting to be in violation of the State's constitution in 2017, and the final status of this voting rule in Maine currently remains uncertain.

If the electorate is supportive of taking the step to use some form of IRV, it turns out that there are some very attractive alternatives that use different criteria to eliminate candidates. For example, the Borda Elimination Rule (BER) eliminates the BR loser in the first step, and this voting rule has some very nice properties. Fishburn and Gehrlein (1976) show that, when $n \to \infty$, BER is the only elimination procedure that always elects the CW, when there is one. In addition, BER can perform better than other elimination rules like PER and NPER at avoiding monotonicity failures. Lepelley et al. (2015) show that, while BER does not minimize the probability of monotonicity failures for $WSR(\lambda)$-based elimination rules in general with IAC, it is the only elimination rule of this type that is completely immune to monotonicity failure in three-candidate elections when a CW exists. As a result, BER presents itself as a very attractive alternative to IRV, particularly because it requires no additional input from voters and it has no additional cost of implementation compared to IRV.

References

Black, D. (1958). *The theory of committees and elections*. Cambridge: Cambridge University Press.

de Condorcet, M. (1788a). On discovering the plurality will in an election. In I. McLean & F. Hewitt (Eds.), *Condorcet: Foundations of social choice and political theory* (pp. 148–156). Hants: Edward Elgar.

de Condorcet, M. (1788b). On the form of elections. In I. McLean & F. Hewitt (Eds.), *Condorcet: Foundations of social choice and political theory* (pp. 139–147). Hants: Edward Elgar.

de Condorcet, M. (1789). On the form of elections. In I. McLean & F. Hewitt (Eds.), *Condorcet: Foundations of social choice and political theory* (pp. 169–189). Hants: Edward Elgar.

Fishburn, P. C., & Gehrlein, W. V. (1976). Borda's rule, positional voting, and Condorcet's simple majority principle. *Public Choice, 28*, 79–88.

Gehrlein, W. V., & Lepelley, D. (2011). *Voting paradoxes and group coherence: The Condorcet efficiency of voting rules*. Berlin: Springer.

Gehrlein, W. V., & Lepelley, D. (2016). Refining measures of group mutual coherence. *Quality and Quantity, 50*, 1845–1870.

Gehrlein, W. V., & Plassmann, F. (2014). A comparison of theoretical and empirical evaluations of the Borda Compromise. *Social Choice and Welfare, 43*, 747–772.

Gehrlein, W. V., Lepelley, D., & Smaoui, H. (2011). The Condorcet efficiency of voting rules with mutually coherent voter preferences: A Borda compromise. *Annals of Economics and Statistics, 101/102*, 107–125.

Gehrlein, W. V., Lepelley, D., & Plassmann, F. (2017, forthcoming). An evaluation of the benefit of using two-stage election procedures. *Homo Oeconomicus*.

Lepelley, D., Moyouwou, I., & Samoui, H. (2015). Monotonicity paradoxes in three-candidate elections using scoring elimination rules. *Social Choice and Welfare*. doi:10.13140/RG.2.1.5075.4961.

Miller, N. R. (2016). *Monotonicity failure in IRV elections with three candidates: Closeness matters*. Working paper. Available at http://userpages.umbc.edu/~nmiller/

Saari, D. G. (1989). A dictionary for voter paradoxes. *Journal of Economic Theory, 48*, 443–475.

Saari, D. G. (1990). The Borda dictionary. *Social Choice and Welfare, 7*, 279–317.

Saari, D. G. (1996). The mathematical symmetry of choosing. *Mathematica Japonica, 44*, 183–200.

Saari, D. G. (1999). Explaining all three-alternative voting outcomes. *Journal of Economic Theory, 87*, 313–355.

Santucci, J. (2016, October 13). Will ranked-choice voting succeed in Maine? That depends on the Democrats. *Washington Post*. Available at https://www.washingtonpost.com/news/monkey-cage/wp/2016/10/13/will-ranked-choice-voting-succeed-in-maine-that-depends-on-the-democrats/?utm_term=.b638c7d53918

Seelye, K. Q. (2016, December 3). Maine adopts ranked-choice voting: What is it and how will it work. *New York Times*. Available at https://www.nytimes.com/2016/12/03/us/maine-ranked-choice-voting.html?_r=0

Wright, S. G., & Riker, W. H. (1989). Plurality and runoff systems and numbers of candidates. *Public Choice, 60*, 155–175.

Young, H. P. (1974). An axiomatization of Borda rule. *Journal of Economic Theory, 9*, 43–52.

Chapter 6
The Impact of Voter Indifference

6.1 Introduction

The analysis that has been performed so far in this study started with the preliminary discussion in Chap. 1 of some voting paradoxes that might be observed in election outcomes with three candidates $\{A, B, C\}$. The descriptions of these paradoxes all began with a consideration of the structure of the preferences that individual voters might have on those candidates. These individual voter preferences were defined by letting $A \succ B$ denote the fact that a voter prefers Candidate A to Candidate B. A voter has complete preferences if there is a preference on each of the possible pairs, so that either $A \succ B$ or $B \succ A$ for all pairs of candidates like A and B, so that no voter indifference is permitted between candidates. It was also assumed that no preference cycles could exist to create intransitivity. Individual voter preferences on candidates that are complete and transitive are defined as linear preference rankings, and the six possible linear preference rankings that each voter might have for three-candidate elections are shown in Fig. 6.1.

It has been assumed to this point that all voters have linear preference rankings on the candidates. The earlier discussion regarding the irrationality of intransitive individual preferences leads us to continue our analysis with the assumption of transitivity, but it is important to turn our attention to the possibility that voter indifference between candidates could exist. This consideration will be found to greatly complicate any analysis that is associated with it.

The possibility that a voter has complete indifference between all three candidates is still ignored, since the resulting lack of any preference for such a voter would have absolutely no impact on any evaluation of how well a voting rule is performing. Six possible cases of partial indifference between three candidates remain to represent the preferences for voters, as shown in Fig. 6.2

The use of the notation $A \sim B$ in Fig. 6.2 indicates that an associated voter is indifferent between any selection of a winner from the pair of Candidates A and B, and a given voting situation that allows for the existence of such partial indifference

W.V. Gehrlein, D. Lepelley, *Elections, Voting Rules and Paradoxical Outcomes*, Studies in Choice and Welfare, DOI 10.1007/978-3-319-64659-6_6

A	A	B	C	B	C
B	C	A	A	C	B
C	B	C	B	A	A
n_1	n_2	n_3	n_4	n_5	n_6

Fig. 6.1 The six possible linear preference rankings on three candidates

$A \sim B$	$A \sim C$	$B \sim C$	A	B	C
C	B	A	$B \sim C$	$A \sim C$	$A \sim B$
n_7	n_8	n_9	n_{10}	n_{11}	n_{12}

Fig. 6.2 Possible preference orderings with partial indifference

for n voters is now defined by a specific set of voter preferences with $n = \sum_{i=1}^{12} n_i$. The voter preferences on candidates that are represented in Fig. 6.2 are all *dichotomous preferences* for which the candidates can be partitioned into a more-preferred subset and a less-preferred subset for each voter. The candidates within each of these two subsets are then indifferent to all other candidates that are included in the same subset, and every candidate in the more-preferred subset is preferred to every candidate in the less-preferred subset.

Candidate A was originally defined as the CW when only linear preference rankings were permitted to represent voters' preferences when:

$$n_1 + n_2 + n_4 > n_3 + n_5 + n_6 \ [\boldsymbol{AMB}]$$
$$n_1 + n_2 + n_3 > n_4 + n_5 + n_6 . [\boldsymbol{AMC}]$$

This definition does not directly accommodate the possibility that voters might have dichotomous preferences, and neither does our original definition of a Weighted Scoring Rule *WSR*(λ). So, what should we do in scenarios when voters might actually have preferences like those in Fig. 6.2?

The simplest and most direct approach to resolving this dilemma is the *Forced Ranking Option* that requires voters to randomly break indifference ties on a pair of candidates and report their resulting linear preference ranking on the candidates, so that a *WSR*(λ) can be used directly in the same fashion that it was used with the original definition. Another approach is to adapt the process for implementing a *WSR*(λ) so that it can directly account for dichotomous preferences. The third possible approach is to consider other voting rules like Approval Voting that are specifically designed to accommodate the presence of indifference in voter preferences, and these three options will be considered in turn

6.2 The Forced Ranking Option

The basic premise of this approach is explained with the following example. Suppose that a voter has dichotomous preferences with Candidate A being more preferred than both B and C, while being completely indifferent to any election outcome that involves B and C. With the Forced Ranking Option, this voter would be required to report that either $A \succ B \succ C$ or $A \succ C \succ B$. If this voter really is indifferent between B and C, it should not make any difference to them which of these two candidates is arbitrarily chosen, since the factor of primary importance to this voter is that A still maintains the status of being the most preferred candidate. And, it does not place very much of a burden on a voter to randomly select one of the two linear rankings in such a scenario. It is then possible to determine the CW and to apply any $WSR(\lambda)$ based on the linear preferences that are reported.

When the voters who have dichotomous preferences like those in Fig. 6.2 are required to randomly break their indifference ties on candidates, there are 12 possible outcomes from the Forced Ranking Option that are displayed in Fig. 6.3.

The mechanism for obtaining the rankings in Fig. 6.3 from the dichotomous preferences in Fig. 6.2 can be explained with the example that there are n_7 voters with $A \sim B$ in the more-preferred subset in Fig. 6.2. And, Fig. 6.3 indicates that n'_7 of these n_7 voters broke the indifference tie with $A \succ B$, while n^*_7 of these voters broke the indifference tie with $B \succ A$. The common linear preference rankings from Figs. 6.1 and 6.3 are not accumulated in the analysis that follows. That is, for example, there are $n_1 + n'_7 + n'_{10}$ linear rankings that all have $A \succ B \succ C$ after the Forced Ranking Option is implemented, but these are not considered to be exactly the same common ranking, since each resulted from a different source. As a result, there are now 18 different linear rankings that must be accounted for in any analysis with this option.

The PMR relationship $\boldsymbol{M}$ that was given above for the original case in which voters are required to have linear preference rankings must now be modified to account for the 18 possible rankings with the Forced Ranking Option, and it is denoted as $\boldsymbol{M}^*$. The *Forced Condorcet Winner*, or *Forced Winner*, is Candidate A based on the preference rankings of voters from Figs. 6.1 and 6.3 when:

A	A	B	A	B	C
B	C	C	B	A	A
C	B	A	C	C	B
n'_7	n'_8	n'_9	n'_{10}	n'_{11}	n'_{12}
B	C	C	A	B	C
A	A	B	C	C	B
C	B	A	B	A	A
n^*_7	n^*_8	n^*_9	n^*_{10}	n^*_{11}	n^*_{12}

Fig. 6.3 Forced ranking options from dichotomous preferences in Fig. 6.2

$$n_1 + n_2 + n_4 + n'_7 + n'_8 + n^*{}_8 + n'_{10} + n^*{}_{10} + n'_{12}$$
$$> n_3 + n_5 + n_6 + n^*{}_7 + n'_9 + n^*{}_9 + n'_{11} + n^*{}_{11} + n^*{}_{12} \qquad [AM^*B]$$

$$n_1 + n_2 + n_3 + n'_7 + n^*{}_7 + n'_8 + n'_{10} + n^*{}_{10} + n'_{11}$$
$$> n_4 + n_5 + n_6 + n^*{}_8 + n'_9 + n^*{}_9 + n^*{}_{11} + n'_{12} + n^*{}_{12}. \qquad [AM^*C]$$

It is obvious from the complexity of these modified relationship that things are indeed becoming significantly more complicated with the addition of the possibility that voters can have dichotomous preferences.

The objective is to consider the probability that a Forced Winner exists, and the associated Condorcet Efficiencies of voting rules, as a function of the number of voters who initially have linear preference rankings, with $k_{LR} = \sum_{i=1}^{6} n_i$ from Fig. 6.1. The same procedure that has been utilized previously for the limiting case of voters as $n \to \infty$ is employed here, and the proportion of voters with linear preference rankings is defined by $\alpha_{LR} = \frac{k_{LR}}{n}$. The proportion of voters with dichotomous preferences obviously follows as $1 - \alpha_{LR}$.

6.2.1 Condorcet Winners with the Forced Ranking Option

The basic IC and IAC models clearly cannot be directly applied when the Forced Ranking Option is considered, so some modifications must be made. The IC model was based on the assumption that the probability vector $\boldsymbol{p}$ that defined the likelihood that each of the six linear preference rankings in Fig. 6.1 were equally likely to represent the preferences of a randomly selected voter, with $p_i = \frac{1}{6}$. An extension of IC was developed in Gehrlein and Valognes (2001) to obtain the probability vector $\boldsymbol{p}^*$ over the 18 rankings in Figs. 6.1 and 6.3, and we refer to that model as the assumption of the *Force Ranking Impartial Culture Condition*, which is denoted by $FIC(\alpha_{LR})$. The underlying notion of this approach is to keep α_{LR} fixed, and then to separately apply the equally likely scenario of IC to the linear preference rankings and to dichotomous rankings. So, when a voter preference profile is being obtained with $FIC(\alpha_{LR})$, it is assumed for each voter that each linear preference ranking in Fig. 6.1 has a probability $\frac{\alpha_{LR}}{6}$ of being selected, and each forced ranking in Fig. 6.3 has a probability $\frac{1-\alpha_{LR}}{12}$ of being selected as that voter's preference.

A representation was obtained for the limiting probability $P_{FW}^{FIC(\alpha_{LR})}(\infty, \boldsymbol{p}^*)$ that a Forced Winner exists with the assumption of $FIC(\alpha_{LR})$ for a specified α_{LR}. It was found that this probability was constant for all α_{LR}, and that it is the same as the limiting probability that a CW exists with IC. Since we found that $P_{CP}^{IC}(\infty, \boldsymbol{p}) = .08774$, it follows that $P_{FW}^{FIC(\alpha_{LR})}(\infty, \boldsymbol{p}^*) = .91226$ for all α_{LR}.

A very similar modification of IAC to account for the Forced Ranking Option is developed in Friese et al. (2016), and it is denoted by $FIAC(\alpha_{LR})$. This approach is a direct variation of the IAC-based conditional distributions that were used to specify particular parameter values of the Measures of Group Mutual Coherence in Chap. 4.

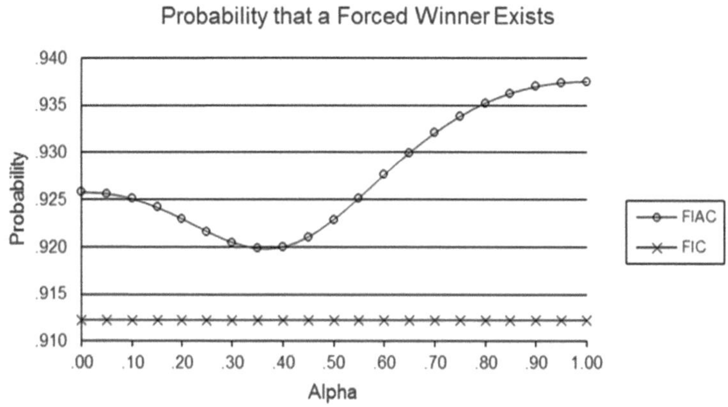

Fig. 6.4 Computed values for the probability that a forced winner exists

That is, it assumes that all voting situations that have a particular value of α_{LR} are equally likely to be observed. As in the earlier applications, this does not imply that all of the values of α_{LR} are equally likely to be observed. A representation was obtained for the limiting probability $P_{FW}^{FIAC(\alpha_{LR})}(\infty)$ that a Forced Winner will exist with $FIAC(\alpha_{LR})$, and it does not have a constant value as we observed with the assumption of $FIC(\alpha_{LR})$. Computed values of $P_{FW}^{FIAC(\alpha_{LR})}(\infty)$ are displayed graphically in Fig. 6.4 over the range of $0 \le \alpha_{LR} \le 1$, along with values of $P_{FW}^{FIC(\alpha_{LR})}(\infty, \boldsymbol{p}^*)$.

These results show that $P_{FW}^{FIAC(\alpha_{LR})}(\infty)$ does not change monotonically as α_{LR} increases, but the degree of dependence among voters' preferences that is inherently introduced by any IAC-based model results in a consistent increase in $P_{FW}^{FIAC(\alpha_{LR})}(\infty)$ compared to the case of complete independence that is present with $P_{FW}^{FIC(\alpha_{LR})}(\infty, \boldsymbol{p}^*)$.

The degree of dependence that has just been introduced with $FIAC(\alpha_{LR})$ relative to $FIC(\alpha_{LR})$ is not the same as the degree of dependence that was introduced earlier with IAC compared to IC, despite the fact that three-candidate elections are being considered in both cases. In the original comparison, IAC probabilities were consistent with a P-E urn model that started with six balls of different colors to represent the six possible linear preference rankings. When the possibility of dichotomous preferences is introduced, the urn model starts out instead with balls of 18 different colors to represent the rankings in Figs. 6.1 and 6.3. Just as in the discussion of the development of Table 2.5, when the impact of having more than three candidates was being considered, the introduction of these additional balls of different colors to start the experiment significantly reduces the probability that consecutively drawn balls will have the same color. However, even with this reduced degree of dependence among voters' preferences with $FIAC(\alpha_{LR})$, the positive impact of that dependence on the probability that a Forced Winner exists is clearly evident in Fig. 6.4.

6.2.2 Condorcet Efficiencies with the Forced Ranking Option

There is a high probability that a Forced Winner will exist, so it is of interest to consider the conditional probability that some common voting rules will elect that candidate, given that such a candidate exists. This probability is referred to as the *Forced Efficiency* of the voting rules. The primary point of interest is to determine the impact that the presence of dichotomous preferences has on these efficiency values. We focused on three single-stage voting rules in our earlier analysis: PR, NPR and BR. The performance of NPR was found to be so poor that it is eliminated from any further consideration, which leaves us to evaluate PR and BR.

Representations for the limiting Forced Efficiency as $n \to \infty$ that a voting rule VR will elect the Forced Winner, given that such a candidate exists, were obtained with the assumption of $FIC(\alpha_{LR})$ in Gehrlein and Valognes (2001). This probability is denoted by $FE_{VR}(\infty, FIC(\alpha_{LR}))$, and mirroring the results that were obtained for the probability that a Forced Winner exists, this Forced Efficiency is constant for all α_{LR} with each single-stage voting rule. Moreover, these Forced Efficiencies are identical to the limiting Condorcet Efficiency for each voting rule with IC.

Representations for the limiting Forced Efficiency for PR and BR are developed in Friese et al. (2016) for the case of $FIAC(\alpha_{LR})$. Computed values of the Forced Efficiencies with PR and BR are displayed graphically in Fig. 6.5 for the range of $0 \leq \alpha_{LR} \leq 1$, along with values of $FE_{PR}(\infty, FIC(\alpha_{LR}))$ and $FE_{BR}(\infty, FIC(\alpha_{LR}))$.

The obvious conclusion from Fig. 6.5 is that the introduction of some dependence among voters' preferences with $FIAC(\alpha_{LR})$ has a dramatic impact on the Forced Efficiency of PR. A similar increase in the Forced Efficiency also exists with BR, but the results are not as significant. However, any potential increases for BR Efficiency are necessarily quite constrained to begin with, since it started out so high with $FIC(\alpha_{LR})$.

All of these results look very strongly supportive of the policy of using the Forced Ranking Option, particularly if there is some degree of dependence among

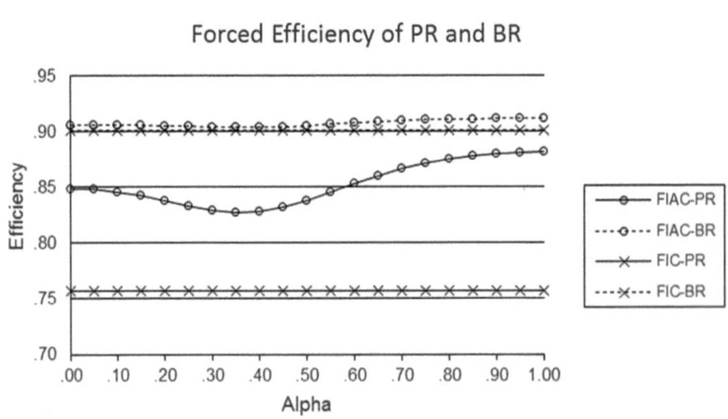

Fig. 6.5 Forced efficiency of PR and BR with $FIC(\alpha_{LR})$ and $FIAC(\alpha_{LR})$

voters' preferences. And, it really does not seem to be an overly burdensome demand for voters to be required to arbitrarily break any indifference ties to provide a linear preference ranking on their ballot. As Captain Edward Smith of the purportedly unsinkable Titanic might have said on the evening of April 14, 1912 when the minor nuisance of some nearby icebergs was first brought to his attention: "What could possibly go wrong with this scenario?"

6.2.3 Significant Problems with the Forced Ranking Option

It is suggested in Gehrlein (2010) that only the actual preferences that are expressed in Figs. 6.1 and 6.2 should be used to determine the *Actual Condorcet Winner*, or *Actual Winner*, to more accurately represent the true preferences of the voters. This requires a modified PMR relationship, $\boldsymbol{M}'$, to replace the $\boldsymbol{M}^*$ relationship from the Forced Ranking Option. Candidate A is then defined as the Actual Winner when:

$$n_1 + n_2 + n_4 + n_8 + n_{10} > n_3 + n_5 + n_6 + n_9 + n_{11}\ [A\boldsymbol{M}'B]$$
$$n_1 + n_2 + n_3 + n_7 + n_{10} > n_4 + n_5 + n_6 + n_9 + n_{12}\ [A\boldsymbol{M}'C].$$

The first indication that something could go very wrong with the Forced Ranking Option results from the fact that the Forced Winner could be a different candidate than the Actual Winner, only as a result of fictitious preferences that were created by the arbitrary breaking of ties for dichotomous preferences with the Forced Ranking Option. So, it suddenly seems like it might not be a very good idea to evaluate voting rules based on their ability to elect the Forced Winner.

The *Extended Impartial Culture Condition*, which is denoted as $EIC(\alpha_{LR})$, was defined in Fishburn and Gehrlein (1980) to develop a representation for the limiting probability that an Actual Winner will exist. It is an IC-based model which adopts the assumption of independence between voters' preferences, and it is defined in a very similar manner as $FIC(\alpha_{LR})$ above. In particular, $\boldsymbol{p}'$ is a probability vector over the 12 rankings in Figs. 6.1 and 6.2. For a specified value of α_{LR}, each linear preference ranking in Fig. 6.1 has a probability $\frac{\alpha_{LR}}{6}$ of representing a voter's preferences and each dichotomous preference ranking in Fig. 6.2 has a probability $\frac{1-\alpha_{LR}}{6}$ of representing a voter's preference.

A limiting representation as $n \to \infty$ for the probability $P_{AW}^{EIC(\alpha_{LR})}(\infty, \boldsymbol{p}')$ that an Actual Winner exists for a specified α_{LR} is obtained as

$$P_{AW}^{EIC(\alpha_{LR})}(\infty, \boldsymbol{p}') = \frac{3}{4} + \frac{3}{2\pi} Sin^{-1}\left(\frac{1}{2+\alpha_{LR}}\right).$$

This representation directly corresponds to the well-known result for the limiting probability $P_{CP}^{IC}(\infty, \boldsymbol{p})$ that Condorcet's Paradox will be observed with IC with $\alpha_{LR} = 1$ from Guilbaud (1952) in Chap. 2. It also verifies another result from Inada

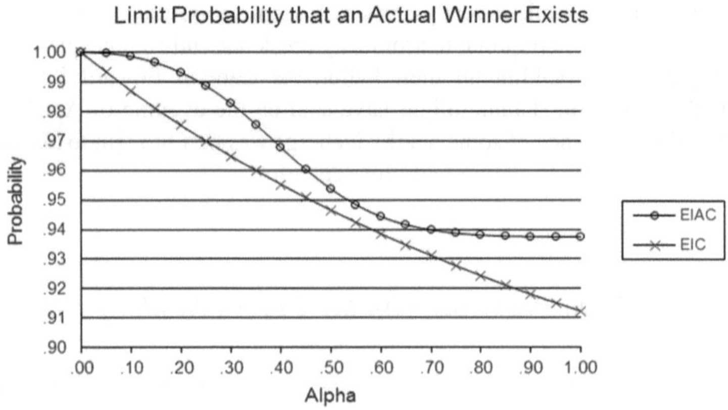

Fig. 6.6 Limit probability that an actual winner exists for $EIC(\alpha_{LR})$ and $EIAC(\alpha_{LR})$

(1964) that an Actual Winner must exist when all voters have dichotomous preferences with $\alpha_{LR} = 0$. Computed values of $P_{AW}^{EIC(\alpha_{LR})}(\infty, \boldsymbol{p}')$ are displayed graphically in Fig. 6.6 for the range of $0 \le \alpha_{LR} \le 1$.

The results in Fig. 6.6 show that the limiting probability that an Actual Winner exists with $EIC(\alpha_{LR})$ consistently decreases as α_{LR} increases, so that an increased proportion of voters with dichotomous preferences has a consistent impact to increase the probability that an Actual Winner will exist. This is a very different outcome than what we observed for the likelihood that a Forced Winner will exist with $FIC(\alpha_{LR})$ in Fig. 6.4.

The same basic concept was adopted by Gehrlein and Lepelley (2015) to account for the possibility of dichotomous preferences with IAC. The *Extended IAC Condition*, which is denoted as $EIAC(\alpha_{LR})$, assumes that all voting situations from Figs. 6.1 and 6.2 that have a specified proportion of voters α_{LR} with linear preference rankings are equally likely to be observed. Since this represents an IAC-based model, it obviously introduces some degree of dependence among voters' preferences. A representation was obtained by Friese et al. (2016) for the associated limiting probability $P_{AW}^{EIAC(\alpha_{LR})}(\infty)$ that an Actual Winner exists, and resulting computed probabilities over the range of $0 \le \alpha_{LR} \le 1$ are shown in Fig. 6.6. The degree of dependence that $EIAC(\alpha_{LR})$ introduces among voters preferences consistently causes an increased likelihood that an Actual Winner will exist, relative to the case of complete independence with $EIC(\alpha_{LR})$. This probability also behaves much more in agreement with expectations, since it decreases monotonically as α_{LR} increases; as opposed to the probability that a Forced Winner exists with $FIAC(\alpha_{LR})$ in Fig. 6.4.

When Actual Winners are considered, it definitely appears that very different results are obtained, compared to outcomes when the Forced Ranking Option is used. And, the logic behind the determination of the Actual Winner makes it a more appealing choice than the Forced Winner. However, the Forced Ranking Option

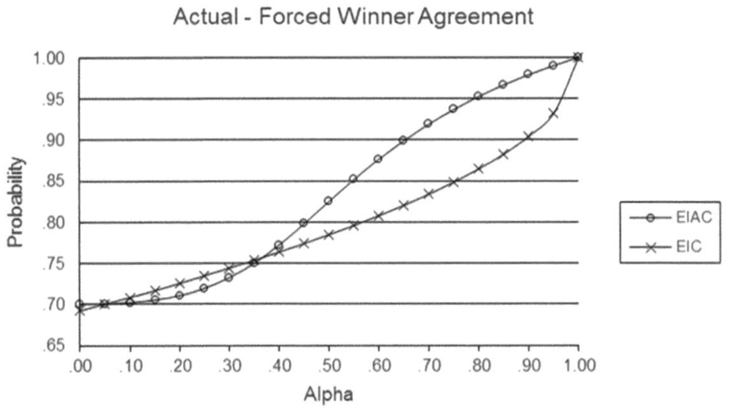

Fig. 6.7 Probability of actual-forced winner agreement with $EIC(\alpha_{LR})$ and $EIAC(\alpha_{LR})$

makes it easy to directly apply the single-stage voting rules that have been considered. So, before we completely reject the Forced Ranking Option, we must first consider how much of a difference really exists between these two scenarios.

The first step to evaluate how significant the differences between the Actual Winner and the Forced Winner really are was performed in Gehrlein (2010). A representation is obtained for the limiting conditional probability that there is mutual agreement on a common candidate with both approaches, given that an Actual Winner exists with $EIC(\alpha_{LR})$. Computed values from that representation are displayed in Fig. 6.7 over the range of $0 \leq \alpha_{LR} \leq 1$.

It is obvious that the Actual and Forced Winners must be the same when all voters have linear preference rankings with $\alpha_{LR} = 1$, so that there are no indifference ties to be broken on any pair of candidates. What is very surprising in Fig. 6.7 is that this probability of mutual agreement consistently decreases to levels below .70 when all voters have dichotomous preferences with $\alpha_{LR} = 0$.

The same type of analysis was performed with $EIAC(\alpha_{LR})$ in Friese et al. (2016), and those results are displayed in Fig. 6.7. While the insertion of a degree of dependence among voters' preferences is found to significantly improve the probability of mutual agreement for the range of $\alpha_{LR} \geq .35$, the same very low levels of mutual agreement still remain for the scenario in which all voters have dichotomous preferences. This is not a good sign to encourage any strong endorsement of the Forced Ranking Option, since the Forced Efficiencies in Fig. 6.5 indicate that PR and BR might have a very good chance of electing the wrong winner when a large proportion of voters have dichotomous preferences with $EIAC(\alpha_{LR})$. The direct impact of this lack of mutual agreement is best displayed with a consideration of the *Actual Efficiency* to measure the propensity of voting rules to elect the Actual Winner with the Forced Ranking Option.

Representations for the limiting Actual Efficiency of PR and BR with the assumption of $FIC(\alpha_{LR})$ were obtained in Gehrlein and Lepelley (2011), and the results are displayed in Fig. 6.8 over the range of $0 \leq \alpha_{LR} \leq 1$.

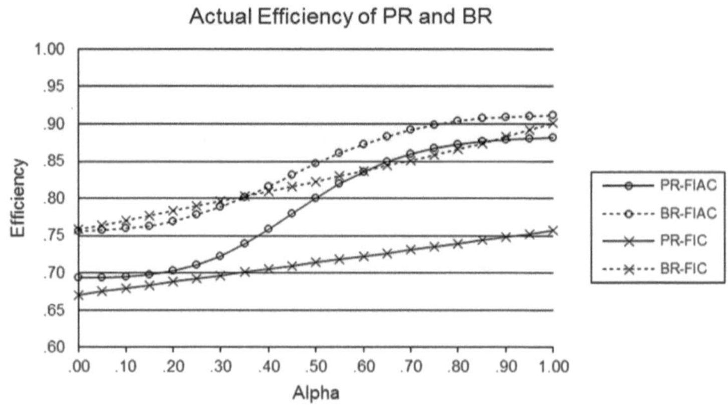

Fig. 6.8 Limiting actual efficiency of PR and BR with $FIC(\alpha_{LR})$ and $FIAC(\alpha_{LR})$

These results show that there is a relatively low probability of only about .67 that the Actual Winner will be elected by PR when all voters' preferences are dichotomous with $\alpha_{LR}=0$. Better performance does exist for BR in the same scenario with an efficiency of about .75, but this still is not an encouraging result for the Forced Ranking Option.

The only remaining hope for endorsing the Forced Ranking Option exists with the possibility that introducing a degree of dependence into voters' preferences with the assumption of $FIAC(\alpha_{LR})$ might result in a more supportive outcome. Friese et al. (2016) developed representations for the limiting Actual Efficiency of PR and BR with $FIAC(\alpha_{LR})$ and the results are shown in Fig. 6.8, where the Actual Efficiency results for PR improve dramatically with larger values of α_{LR}. The BR efficiencies do not improve as much, but they always maintain superior performance relative to PR. However, the associated efficiency values for voting situations with completely dichotomous preferences at $\alpha_{LR}=0$ only increase by a very small amount for both PR and BR, compared to the results with $FIC(\alpha_{LR})$. So, the possibility that the wrong candidate might be elected with the Forced Ranking Option still remains as a significant issue that must be considered problematic, even with the introduction of dependence among voters' preferences.

6.3 Modifying Scoring Rules for Dichotomous Preferences

The Forced Ranking Option clearly has significant problems that are associated with its use when a significant number of voters have dichotomous preferences. So, the only remaining alternative is to find some way to modify the voting rules that have been considered so that they can accommodate voters with such preferences on candidates, and we continue to evaluate some of these approaches while ignoring voters who have complete indifference between all three candidates. The first step of this process is to look at modifications that can be made to all $WSR(\lambda)$ to accomplish this outcome and produce voting rules with enhanced Actual Efficiencies.

The basic logic that forms the foundation of the definition of an Extended Weighted Scoring Rule, which is denoted by $EWSR(\lambda)$, is developed in Gehrlein and Valognes (2001) and Diss et al. (2010). An $EWSR(\lambda)$ operates exactly like a $WSR(\lambda)$ when voters have linear preference rankings like those in Fig. 6.1. However, when voters have dichotomous preferences like those in Fig. 6.2, an $EWSR(\lambda)$ redistributes the total of $1+\lambda$ points from the $WSR(\lambda)$ over the three candidates. When there are two candidates in the more-preferred subset, each is given an average $\frac{1+\lambda}{2}$ points from the most-preferred and middle-ranked candidates with the associated $WSR(\lambda)$, and the candidate in the less-preferred subset still receives zero points as the least-preferred candidate. If there is only one candidate in the more-preferred subset with dichotomous preferences, the $EWSR(\lambda)$ will award it one point as the most-preferred candidate, and the two candidates in the less-preferred subset will each receive the average $\frac{\lambda+0}{2}$ points from the middle-ranked and least-preferred candidate with the associated $WSR(\lambda)$.

The use of an $EWSR(\lambda)$ therefore treats all voters equally, since it consistently allocates the same $1+\lambda$ points to the candidates for each voter, and it does completely eliminate the problems that can result from the imposition of the Forced Ranking Option. However, it does add an increased level of complexity to the voting procedure that must both be explained to and accepted by the electorate. The obvious point of interest is to determine if the use of $EWSR(\lambda)$ significantly improves the Actual Efficiency with the three primary extensions: EPR, ENPR and EBR. Gehrlein and Valognes (2001) develop a limiting representation for the Actual Efficiency of $EWSR(\lambda)$ with the assumption of $EIC(\alpha_{LR})$, and it was found that the associated efficiencies for EPR and ENPR are identical. It was also found that the Actual Efficiency of EBR is one when all voters have dichotomous preferences with $\alpha_{LR}=0$.

Computed values from that representation are shown in Fig. 6.9 for EPR and EBR, where EBR is found to significantly dominate EPR over the entire range of $0 \le \alpha_{LR} \le 1$. Both EPR and EBR have Actual Efficiencies that consistently increase as the proportion of voters with linear preference rankings decreases, so that an increased proportion of voters with dichotomous preferences is favorable to these $EWSR(\lambda)$. These results also represent a very significant improvement over the associated Actual Efficiencies that appear in Fig. 6.8 where the Forced Ranking Option was imposed. This notable improvement is particularly evident in the efficiency results for EBR, which has a value of one when all voters have dichotomous preferences.

This line of analysis was extended to consider the impact that the introduction of some degree of dependence among voters' preferences might have with $EIAC(\alpha_{LR})$ in Gehrlein and Lepelley (2015). Actual Efficiency values from that analysis are displayed graphically in Fig. 6.10 for EPR, ENPR and EBR.

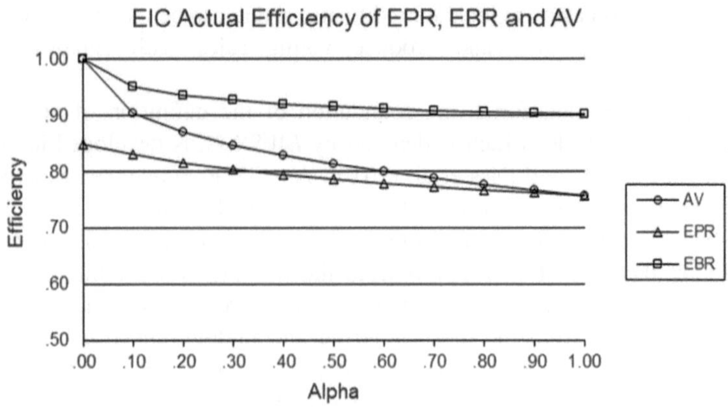

Fig. 6.9 Limiting actual efficiency with $EIC(\alpha_{LR})$

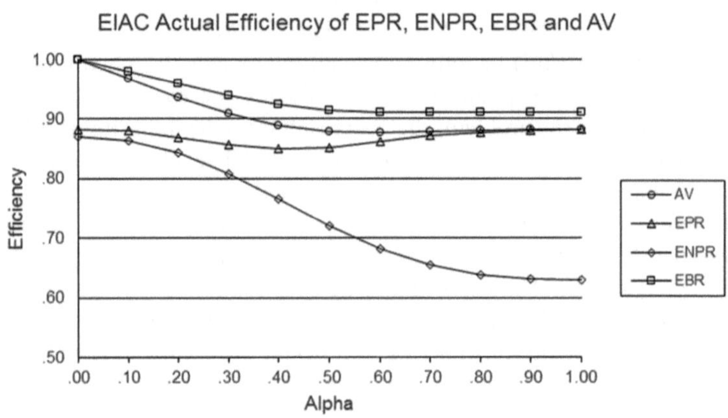

Fig. 6.10 Limiting actual efficiency with $EIAC(\alpha_{LR})$

The impact of introducing some dependence among voter's preferences is very significant in improving the Actual Efficiency for EPR, and there is less of an impact on the EBR efficiencies that were already relatively high. We see the overall pattern that EBR dominates EPR, which in turn significantly dominates ENPR. The use of these $EWSR(\lambda)$ clearly resolves the significant issues that arose when the Forced Ranking Option was imposed, particularly with EBR.

6.4 Approval Voting

Approval Voting (AV) presents a very different approach to dealing with the possible presence of dichotomous voters' preferences, and each voter is permitted to cast a vote to approve of any number of candidates. It would typically be expected that a vote would be cast only for the most preferred candidate for voters with linear preference rankings like those in Fig. 6.1. When voters have dichotomous preferences like those in Fig. 6.2, it is expected that they would cast one vote for each candidate in their more-preferred subset. So, each voter could cast either one or two votes, and we continue to ignore the possibility that any voter would vote for all three candidates. The use of AV therefore abandons the principle of $EWSR(\lambda)$ that all voters should have the same number of points to distribute to candidates. But, all voters are still treated equally with AV, in the sense that they all can vote for as many candidates as they wish.

The formal mathematical analysis of AV began with an unpublished working paper by Weber (1978), and efforts to encourage its use were heavily promoted by Brams and Fishburn (1983). But, it can be very difficult to explore many areas of the study of voting rules without finding traces of Condorcet's footprints that were left behind many years earlier, and the study of AV is not an exception. Condorcet (1788) proposed an election procedure that simplified a number of his previously proposed election rules for electing members to Provincial Assemblies. The rule is very similar to the nature of AV, and it requires each voter to make up a list of the 20 candidates whom they most prefer. The winner is then selected as the candidate who receives the greatest number of votes from these lists. This specified number of candidates to be listed differs from the basic principle of AV. It is further suggested by Condorcet that these lists should provide preference rankings on candidates, to be used if necessary as a tie-breaking mechanism.

Condorcet goes on to describe scenarios in which he believes that such an election rule is particularly appropriate for use. That is, when it is not necessary to elect the most worthy candidate, but when it is sufficient to be certain of electing a candidate who is capable of competently performing the associated duties of a position that "does not require any particular talents" (Condorcet 1788, p. 141). It is important to note that Condorcet is making this assessment in terms of electing one member of an assembly, since he later makes a distinction based on that fact.

Condorcet (1789) presents an extensive discussion about the primary rules that are essential to holding elections properly. It is not at all surprising to find that Rule 11 specifically states that the only rigorous way to find the best candidate is to use "two by two" comparisons to find the CW. However, he notes that this method is very time consuming, complicated and awkward to use. So, in Rule 12 he then states that other rules should be considered, as long as the winner is at least not considered unworthy by a majority of the voters. This is made clear in his argument (Condorcet 1789, pp. 177–178) that it is more important to elect a qualified candidate to an office than it is to expend great effort in an unlikely pursuit of determining the most qualified candidate. However, it is added that this is particularly true "when we are not electing a man to hold office on his own," to indicate that Condorcet was reluctant to extend his endorsement of such a rule to the election of a single person to a position. This stipulation is emphasized when he further gives a criticism of such rules as possibly having a propensity to elect candidates only because most of the electorate is indifferent to them, and warns that (Condorcet 1789, p. 178) "In avoiding bad choices we must not favour mediocre ones."

So, a direct application of AV really might not suit Condorcet when a single candidate is being elected to act alone in a position. Criticism that is currently aimed at AV is typically based on this same concern that Condorcet expressed many years ago. That is, there is often an opinion that AV might favor the election of mediocre candidates, only because there is no strong opposition to them. The validity of this objection to the use of AV can be accurately assessed with the measure of its Condorcet Efficiency.

6.4.1 Condorcet Efficiency of Approval Voting

A preliminary attempt to assess the Condorcet Efficiency of AV was performed in Gehrlein and Lepelley (1998), where all voters are assumed to have linear preference rankings for the general case of m candidates in the limiting case of voters as $n \rightarrow \infty$. A fixed probability vector $\boldsymbol{h}$ is defined to establish the likelihood of the number of candidates that each voter would approve of with AV. The outcome of the IC-based analysis with this model is disheartening for supporters of AV, since it finds that the AV Efficiency is identical to the efficiencies of both PR and NPR for any $\boldsymbol{h}$, so that AV does not gain any advantage over these two voting rules that are known to be weak performers. But, there is something else that must be considered. What happens if the possibility of dichotomous voters' preferences is directly added to this mix, and we then consider the Actual Efficiency of AV?

Diss et al. (2010) and Gehrlein and Lepelley (2015) use the assumption of $EIC(\alpha_{LR})$ to develop representations for the limiting Actual Efficiency of AV, and they are identical for all $\boldsymbol{h}$. Computed values from those results are displayed graphically in Fig. 6.9 for the range $0 \leq \alpha_{LR} \leq 1$, and we see the obvious fact that AV has the same efficiency as EPR when all voters have linear preference rankings with $\alpha_{LR} = 1$. The AV Efficiency continuously increases as α_{LR} decreases to represent an increased proportion of voters with dichotomous preferences, and it is maximized at one when all voters have dichotomous preferences. The presence of an increased proportion of voters with dichotomous voters is even more favorable to AV Efficiency than it is to EBR Efficiency, but AV still never dominates EBR, because it starts out at such a low level when all voters have linear preference rankings.

The impact that the addition of a degree of dependence among voters' preferences with $EIAC(\alpha_{LR})$ has on AV Efficiency is considered in Gehrlein and Lepelley (2015), where it is assumed that voters with linear preference rankings will only vote for one candidate. The limiting Actual Efficiency values from representations that are developed in that study are shown in Fig. 6.10, and the trends from that analysis are similar to those observed in Fig. 6.9 with $EIC(\alpha_{LR})$. The presence of dependence among voters' preference causes higher efficiencies for EPR and EBR to start out with when all voters have linear preference rankings, so the magnitude of the impact of increasing the proportion of voters with dichotomous preferences is not as striking in this case.

All of this strongly reinforces the conclusion of Diss et al. (2010) that the introduction of indifference into voters' preferences will give an Actual Efficiency advantage to AV over both EPR and ENPR. However, EBR is found to dominate AV in all cases for both $EIC(\alpha_{LR})$ and $EIAC(\alpha_{LR})$, except when all voters have dichotomous preferences, where AV and EBR both elect the Actual Winner with certainty. The presence of a large proportion of voters with dichotomous preferences in voting situations therefore does not support the assertion that AV has a propensity to elect mediocre candidates. However, the same level of support is not as clearly present for AV when the proportion of voters with dichotomous preferences is small.

6.4.2 *Strong Borda Paradox Probabilities with Extended Rules*

The fact that the Actual Efficiency of AV is near one when a large proportion of voters have dichotomous preferences indicates that it is very unlikely to elect the Actual Condorcet Loser. The probability of observing a Strong Borda Paradox, as described in Chap. 1, must therefore be small in such cases, to make AV look quite appealing. Gehrlein and Lepelley (2017) obtained representations for the limit probability that a Strong Borda Paradox is observed by both AV and $EWSR(\lambda)$ with the assumption of $EIC(\alpha_{LR})$. This probability was found to be identical for EPR and ENPR. Computed values from those representations are displayed in Fig. 6.11 for AV, EPR and EBR over the range $0 \leq \alpha_{LR} \leq 1$.

These results show AV with the same probability of exhibiting a Strong Borda Paradox as EPR when all voters have linear preference rankings. As α_{LR} decreases, this probability consistently decreases for AV, and it predictably has a value of zero when all voters have dichotomous preferences.

Gehrlein et al. (2016) and Gehrlein and Lepelley (2017) extended this same type of analysis to consider the impact that introducing dependence among voters' preferences has on the probability that a Strong Borda Paradox is observed by using the assumption of $EIAC(\alpha_{LR})$. Computed values from the representations that are obtained are shown in Fig. 6.12.

The introduction of dichotomous preferences into voters' preferences consistently shows that AV has a reduced propensity to exhibit a Strong Borda Paradox compared to both EPR and ENPR in all cases for both $EIC(\alpha_{LR})$ and $EIAC(\alpha_{LR})$.

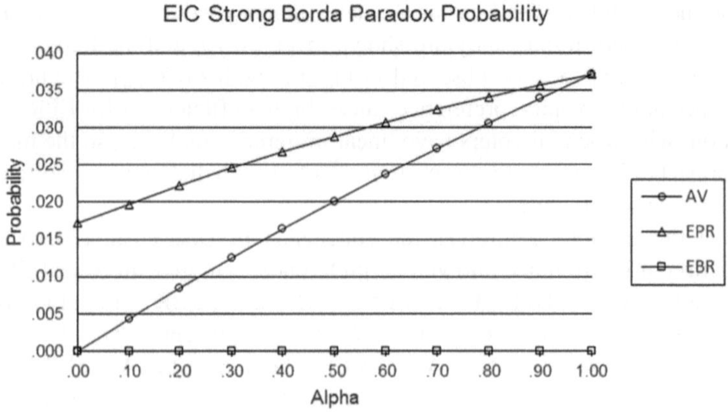

Fig. 6.11 Limiting strong Borda Paradox probability with $EIC(\alpha_{LR})$

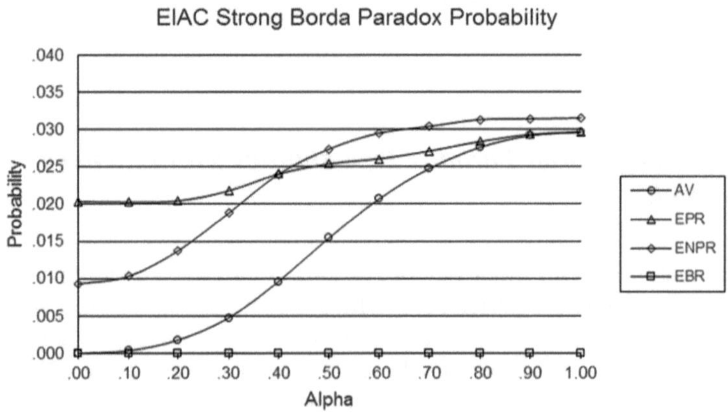

Fig. 6.12 Limiting strong Borda Paradox probability with $EIAC(\alpha_{LR})$

However, EBR is found to be completely immune to exhibiting this paradox, to give it a dominance over AV under this criterion.

6.4.3 Approval Voting: Other Factors of Evaluation

It was mentioned previously that the primary early promoters of AV were Steven Brams and Peter Fishburn, and we discuss all of this in some detail because AV has attracted many supporters. Brams and Fishburn (1983) compiled a list of benefits that could be expected from using AV. This list is largely focused on how AV would be an improvement over PR, which would presumably be implemented with the Forced Ranking Option:

1. It allows voters the flexibility to support a candidate that they favor, without "wasting a vote."
2. It encourages voter turnout, since support can be shown for candidates with a small possibility of winning, and voters are not forced to decide between ties for their most-preferred candidate.
3. It helps to elect the candidate with the most overall support, so it induces candidates to establish platforms that are acceptable to a majority of voters.
4. Minority candidates receive proper recognition, since voters can support them without having to transfer a single PR vote to a less preferred candidate that has some chance of winning.
5. It is insensitive to the number of candidates, unlike PR where a winner can be selected with only a small percentage of the total votes when there are many candidates.
6. It prevents situations in which candidates can enter a race to 'splinter' the electorate to win by PR with a small percentage of the total vote.
7. It would be very easy to implement AV as a voting system.
8. It has a strong propensity to elect the CW.

This final claim definitely must be considered in light of what we have learned above, since the Actual Efficiency of AV is highly dependent on the proportion of voters with dichotomous preferences.

A list was also created to highlight some potential difficulties that could arise from using AV:

9. It overlooks some refinements in voter preferences that ranking systems could detect, but it is asserted that AV will likely produce the same winner as a ranking procedure anyway.
10. It could encourage candidates to adopt uncertain positions on issues, to possibly appeal to a broader base of voters.
11. It could undermine the "two-party system", but the major parties would still continue to exist if they maintain positions that appeal to a significant voter base.
12. It creates inequities among voters, based on the number of candidates that they vote for. But, AV is more equitable in terms of allowing voters to express their preferences more adequately.

Brams and Fishburn (1978) established a model that initially specifies some restrictions on the allowable forms of preferences that voters might have on combinations of candidates. A formal definition of admissible voting behavior is then specified when these allowable forms represent voters' preferences. With these restrictions and specifications, AV is then shown to have many superior qualities relative to other voting rules when all voters have dichotomous preferences, to the degree that AV is even shown to be immune to strategic manipulation. All of this perfectly coincides with our observation that the performance of AV is very strongly linked to the proportion of voters that have dichotomous preferences.

Soon after the analysis of Brams and Fishburn (1983) was presented, Niemi (1984) produced examples to show that very different outcomes are possible with AV as soon as the assumption that voters have dichotomous preferences is dropped. Since the dominant performance of AV is so heavily dependent on the assumption that voters have dichotomous preferences, it is clearly of interest to consider the efficacy of this assumption. As a result, Radcliff (1993) performs an empirical study to evaluate the propensity of voters to have dichotomous preferences, based on a model that generates preference rankings for voters from thermometer score responses to surveys regarding US Presidential candidates in American National Election Studies from 1972 to 1984. The results suggest that only approximately 30% of respondents had dichotomous preferences in three-candidate elections, with that percentage decreasing dramatically as the number of candidates increases to four or five. When we use this estimate that $\alpha_{LR} = .70$ with results for $EIC(\alpha_{LR})$ in Fig. 6.9 and $EIAC(\alpha_{LR})$ in Fig. 6.10, very little difference is observed between the Actual Efficiencies of AV and EPR.

Another study by Arrington and Brenner (1984) concludes that many of the superior qualities that are attributed to AV will not be valid in practice. Their argument is primarily based on the belief that the use of AV will encourage more candidates to enter into elections, thinking that they might have a better chance of winning. It is hypothesized that voters' preferences are less likely to be dichotomous as more candidates enter into the election, so that all of the proofs that AV is superior when it is assumed that voters' preferences are dichotomous will tend to become irrelevant. It suffices to say that there has been significant debate regarding the desirability of using AV, with strong opinions being held on both sides. It was mentioned previously that Donald Saari is the strongest promoter of BR, and he is also a staunch critic of AV [See for example Saari and Van Newenhizen (1988)]. An excellent survey of work that is related to AV is presented in *Handbook on Approval Voting* (Laslier and Sanver 2010), where the Introduction of that book contains interesting remarks that are made in evaluating the more negative views of AV that are presented in the work of Nurmi and Saari to the more positive views of Brams and Fishburn. Brams and Fishburn (2005) describe the difficulties that can be encountered in attempts to have a new voting rule like AV adopted for use in actual elections. Given the diversity of opinions regarding AV, it is clearly of interest to consider what impact the use of AV would have had on actual elections.

6.4.4 *Effectiveness of Approval Voting: Empirical Studies*

Most empirical studies have used election results as a basis of models to evaluate how successful AV would have been at showing the true level of support that is present for candidates that do not win by using PR with the Forced Ranking Option, and on the propensity of AV to elect the CW. A survey of a number of these studies that considered US Presidential elections is given in Gehrlein and Lepelley (2011), where it is generally concluded that AV results would definitely give a more accurate reflection of the true level of support for minority candidates who did not win, but that AV would have been unlikely to elect a different candidate than the one selected by PR.

Two very important studies to evaluate AV were conducted in conjunction with the first stage of the two-stage procedure to elect for the President of France. The first step of this process uses PR elimination to reduce the typically large number of candidates down to two candidates who are passed to the second stage. The first study by Laslier and Van der Straeten (2003) performed a very extensive study of the French Presidential election of 2002, in which 16 candidates were competing. The authors received permission to establish a secondary voting station at the actual voting sites during the first-stage elections in two towns. Voters were aware that this second voting station was being run on an experimental basis, and more than 75% of them agreed to vote with AV in the secondary stage in both cases.

Several important observations were made from this experiment that clearly reflects the actual voting situation. First, the average number of candidates that voters actually voted for with AV was 3.15 candidates of the 16 possible candidates. Only a small proportion of voters chose to vote for more than five candidates. Second, the impact that AV would have had on this election is shown by the fact that there would very likely have been a significant change in the candidates who were passed to the second stage. Candidate Le Pen, who represented the extreme right of the political spectrum, had a hard core of supporters, but the total number of other voters who approved of Le Pen was relatively small. Candidate Jospin, who represented moderates in the political spectrum, received a smaller percent of the PR vote than Le Pen, but received a substantially larger number of votes by AV. As a result, Jospin would likely have replaced Le Pen in the runoff election in the second stage, which Le Pen ultimately lost by a wide margin to Jacques Chirac.

The second study by Baujard and Igersheim (2009) reports on a similar experiment with AV during the French Presidential election of 2007. The PR rankings in the actual election results for the towns where testing was done were consistent with the national percentages of PR vote results from all of France. The three candidates with the largest PR vote percentages were Sarkozy (31.2%), Royal (25.9%) and Bayrou (18.8%). However, the results from the follow-up AV election indicate that the percentages for these three candidates completely reversed the PR ranking, with Bayrou (42.8%), Royal (41.6) and Sarkozy (35.9%). Further analysis suggested that Bayrou, who is a political centrist, was the CW. So, AV would have elected Bayrou, while the PER process failed to even pass him to the second stage of the election.

Some concerns were expressed earlier that voting rules like AV might have a tendency to elect mediocre candidates. No general support for this observation has been found in these empirical studies. However, when a difference is found to exist between the outcomes with AV and the standard PR elimination procedures, there is support for the belief that the use of AV is more likely to produce winners who tend to represent the center of the political spectrum, as indicated by the analysis of the preceding example with François Bayrou.

References

Arrington, T. S., & Brenner, S. (1984). Another look at approval voting. *Polity, 17*, 118–134.

Baujard, A., & Igersheim, H. (2009). Expérimentation du vote par note et du vote par approbation le 22 avril 2007: Premiers résultats. *Revue Economique, 60*, 189–201.

Brams, S. J., & Fishburn, P. C. (1978). Approval voting. *American Political Science Review, 72*, 831–847.

Brams, S. J., & Fishburn, P. C. (1983). *Approval voting*. Boston: Birkhäuser.

Brams, S. J., & Fishburn, P. C. (2005). Going from theory to practice: The mixed success of approval voting. *Social Choice and Welfare, 25*, 457–474.

de Condorcet, M. (1788). On the constitution and the functions of provincial assemblies. In I. McLean & F. Hewitt (Eds.), *Condorcet: Foundations of social choice and political theory* (pp. 139–147). Hants: Edward Elgar.

de Condorcet, M. (1789). On the form of good elections. In I. McLean & F. Hewitt (Eds.), *Condorcet: Foundations of social choice and political theory* (pp. 169–189). Hants: Edward Elgar.

Diss, M., Merlin, V., & Valognes, F. (2010). On the Condorcet efficiency of approval voting and extended scoring rules for three alternatives. In J.-F. Laslier & R. M. Sanver (Eds.), *Handbook on approval voting* (pp. 255–283). Berlin: Springer.

Fishburn, P. C., & Gehrlein, W. V. (1980). The paradox of voting: Effects of individual indifference and intransitivity. *Journal of Public Economics, 14*, 83–94.

Friese, E., Gehrlein, W. V., Lepelley, D., & Schürmann, A. (2016). The impact of dependence among voters' preferences with partial indifference. *Quality and Quantity*. doi:10.1007/s11135-016-0446-7.

Gehrlein, W. V. (2010). The impact of forcing preference rankings when indifference exists. In A. Van Deemen & A. Rusinowska (Eds.), *Collective decision making: Views from social choice and game theory* (pp. 17–29). Berlin: Springer.

Gehrlein, W. V., & Lepelley, D. (1998). The Condorcet Efficiency of approval voting and the probability of electing the Condorcet loser. *Journal of Mathematical Economics, 29*, 271–283.

Gehrlein, W. V., & Lepelley, D. (2011). *Voting paradoxes and group coherence: The Condorcet Efficiency of voting rules*. Berlin: Springer.

Gehrlein, W. V., & Lepelley, D. (2015). The Condorcet Efficiency advantage that voter indifference gives to approval voting over some other voting rules. *Group Decision and Negotiation, 24*, 243–269.

Gehrlein, W. V., & Lepelley, D. (2017). *The limiting probability of a Borda Paradox*. Technical Note at: www.researchgate.net/publication/314390895_The_Limiting_Probability_of_a_Strong_Borda_Paradox

Gehrlein, W. V., & Valognes, F. (2001). Condorcet efficiency: A preference for indifference. *Social Choice and Welfare, 18*, 193–205.

Gehrlein, W. V., Lepelley, D., & Moyouwou, I. (2016). A note on approval voting and electing the Condorcet loser. *Mathematical Social Sciences, 80*, 115–122.

Guilbaud, G. T. (1952). Les théories de l'intérêt général et le problème logique de l'agrégation. *Economie Appliquée, 5*, 501–584.

Inada, K. (1964). A note on simple majority decision rule. *Econometrica, 32*, 525–531.

Laslier, J. F., & Sanver, M. R. (2010). *Handbook on approval voting*. Berlin: Springer.

Laslier, J. F., & Van der Straeten, K. (2003). *Approval voting: An experiment during the French 2002 presidential election*. Presented at Third International Conference on Logic, Game Theory and Social Choice, September 2003, Siena, Italy (pp. 294–297).

Niemi, R. G. (1984). The problem of strategic behavior under approval voting. *American Political Science Review, 78*, 952–958.

Radcliff, B. (1993). The structure of voter preferences. *Journal of Politics, 55*, 714–719.

Saari, D. G., & Van Newenhizen, J. (1988). The problem of indeterminacy in approval, multiple and truncated voting systems. *Public Choice, 59*, 101–120.

Weber, R. J. (1978). *Multiply weighted voting systems*. Yale University, unpublished manuscript.

Chapter 7
Other Voting Rules and Considerations

7.1 Introduction

This final chapter is used to consider some additional important issues that are related to the conclusion to this point that BR clearly has an advantage over the other voting rules when all factors are considered. It was mentioned in Chap. 5 that a major criticism of BR is that it is generally perceived to be very susceptible to manipulation, such that a group of voters can misrepresent their true preferences with strategic voting in order to obtain a more preferred outcome from an election. We begin this chapter by providing a thorough analysis of this phenomenon, to see if this common belief is valid. The next section will then consider the Three-valued Scale Evaluative Voting Procedure that is a new voting rule that has been receiving attention as an extension of AV or as a particular case of Range Voting (RV), where RV sets a fixed range of allowable scores and each voter then selects a score to assign to each candidate from that range. The third section will then consider the overall validity of the conclusion that BR generally displays superior performance when attention is changed to consider the case of more than three candidates. The final section will consider the impact that voter abstention rates might have on our conclusions.

7.2 BR and Strategic Voting

The concept of strategic voting, or manipulation, can be demonstrated by returning back to the story of the two friends in Chap. 1, when a voting situation was given in Fig. 1.8 to lead to the No Show Paradox. Both of the friends had the identical preference rankings on three candidates $\{A, B, C\}$ with $A \succ B \succ C$, and this has not changed. When both participated in an election with NPER, their middle-ranked Candidate B was elected, and it was then discovered that their most-favored

W.V. Gehrlein, D. Lepelley, *Elections, Voting Rules and Paradoxical Outcomes*, Studies in Choice and Welfare, DOI 10.1007/978-3-319-64659-6_7

Fig. 7.1 An example preliminary voting situation for strategic voting

A	A	B	C	B	C
B	C	A	A	C	B
C	B	C	B	A	A
$n_1 = 6$	$n_2 = 6$	$n_3 = 7$	$n_4 = 5$	$n_5 = 6$	$n_6 = 5$

Candidate A would have been elected if they had not participated at all in the election.

The story continues, and Candidate B is now nearing the end of that first term, and an effort is being made for re-election. The same three candidates have entered into the race again. But, after the disappointing outcome of the first election, the two friends led a successful campaign to change the method of voting to BR. Two members have been added to the electorate, so there are now 35 voters participating in the election. The preferences of the original voters have also shifted from those in Fig. 1.8, due to the performance of Candidate B during the first term, and the voting situation that represents the current preference rankings of the electorate is displayed in Fig. 7.1.

Without actually knowing the specific details of this voting situation, one of the two friends secretly confesses to having taken the extremely wise step of studying the fascinating topic of election procedures after the awful fiasco of the first election. A general observation regarding the current election scenario is then presented by this wise friend; that preliminary polls and discussions make it look like a close race between A and B, with C having much less of a chance to win. Then, since both know that their true preferences are $A \succ B \succ C$, the suggestion is proposed that both should instead misrepresent their preference ranking as $A \succ C \succ B$, without discussing this questionable strategy with any other voters. The idea behind this strategy is that this will increase the relative level of support for their preferred Candidate A over B, by decreasing the BR score that B will obtain. There is nothing that these two voters can do to increase the actual BR score for A. While this maneuver will simultaneously show more support for their least-preferred Candidate C, that candidate is not considered to be a serious threat to A to begin with. So, what really happens as a result of this devious action by these two conniving friends to misrepresent their preferences with strategic voting?

Based on the preliminary voting situation in Fig. 7.1, we find that Candidate B is the CW, with *BMA* (18-17) and *BMC* (19-16) from the true preferences. When BR is used to determine the winner, we find

$$Score(A) = 1(12) + \frac{1}{2}(12) + 0(11) = 18,$$

$$Score(B) = 1(13) + \frac{1}{2}(11) + 0(11) = 18.5,$$

$$Score(C) = 1(10) + \frac{1}{2}(12) + 0(13) = 16.$$

Fig. 7.2 An example modified voting situation with strategic voting

A	A	B	C	B	C
B	C	A	A	C	B
C	B	C	B	A	A
$n_1 = 4$	$n_2 = 8$	$n_3 = 7$	$n_4 = 5$	$n_5 = 6$	$n_6 = 5$

The CW is therefore elected by BR for this voting situation in the absence of any misrepresented preferences.

When the two friends change their preferences from $A \succ B \succ C$ to $A \succ C \succ B$, the voting situation is modified and the result is shown in Fig. 7.2.

As a result of this misrepresentation of preferences, the BR results become:

$$Score(A) = 1(12) + \frac{1}{2}(12) + 0(11) = 18,$$

$$Score(B) = 1(13) + \frac{1}{2}(9) + 0(13) = 17.5,$$

$$Score(C) = 1(10) + \frac{1}{2}(14) + 0(11) = 17.$$

Candidate A has therefore become the winner with BR, so the two conniving friends with the preference ranking $A \succ B \succ C$ have successfully manipulated BR to obtain a better outcome for themselves by misrepresenting their preferences to vote strategically. This outcome might provide some solace to them after being victimized by the No-Show Paradox in the first election.

However, it is extremely important to recognize that successful manipulation strategies require a lot of thought and careful coordination for things to work as planned. For example, suppose that the two conniving friends had instead decided to carry this deception further, to determine that there are four other voters with the same preference ranking $A \succ B \succ C$, and then convinced all of them to join into a coalition to misrepresent their preferences as $A \succ C \succ B$. The resulting over-modified voting situation that would result from this scenario is shown in Fig. 7.3.

The resulting candidate scores with BR for the over-modified scenario from Fig. 7.3 are given by:

$$Score(A) = 1(12) + \frac{1}{2}(12) + 0(11) = 18,$$

$$Score(B) = 1(13) + \frac{1}{2}(5) + 0(17) = 15.5,$$

$$Score(C) = 1(10) + \frac{1}{2}(18) + 0(7) = 19.$$

By carrying the use of strategic voting too far, some level of karma would be prevailing in this case; since things have backfired very badly on the coalition. The ultimate winner of the election with BR would now be the least-preferred Candidate C for every member of this conniving coalition of six ultimately less-than-strategic

Fig. 7.3 An example over-modified voting situation with strategic voting

A	A	B	C	B	C
B	C	A	A	C	B
C	B	C	B	A	A
$n_1 = 0$	$n_2 = 12$	$n_3 = 7$	$n_4 = 5$	$n_5 = 6$	$n_6 = 5$

voters. It is very clear that successful manipulation would require a significant level of coordination among voters.

The fact that BR can be manipulated is far from being a recent observation. It was mentioned previously that BR was implemented by the French Academy of Sciences sometime after Borda presented his paper on the topic in 1770 until 1801. Many instances of strategic voting were observed, and BR was strongly criticized for this flaw. Borda responded wryly to this criticism by stating that (Black 1958, p. 238) "My scheme is intended only for honest men." It is very interesting to note that the early criticism of this flaw with BR apparently completely overlooked the fact that this same flaw existed with all other voting rules as well. It was not until Gibbard (1973) and Satterthwaite (1975) that it became known that effectively every voting rules is susceptible to strategic voting with three or more candidates. So, to what extent can the alleged high vulnerability of BR to strategic voting be considered to be true when it is compared to other voting rules?

Before we proceed to consider the likelihood that various voting rules are vulnerable to manipulation, a few definitions are needed. *Individual Manipulation* refers to scenarios in which a single voter can change the outcome of an election with strategic voting. When a group of voters is required to accomplish this outcome, it is referred to as *Coalitional Manipulation*. The example that was described above when two friends were successfully using strategic voting is referred to as a scenario that displays *Naïve Behavior* on the part of voters who are not involved in the manipulation, since none of them make any effort to react to the possibility that such strategic voting might be used to their detriment.

7.2.1 Manipulation with Naïve Voters: The Evidence Against BR

A seminal paper in the study of the propensity of PR, BR, PER and NPER to be susceptible to Coalitional Manipulation with naïve voters was performed in Chamberlin (1985) for three-candidate elections. This was a Monte Carlo simulation based analysis that generated random voters' preferences from two different models: with the basic IC scenario and with a geometric spatial model of voter preferences. These results generally indicate that PER tends to be the voting rule that is least susceptible to manipulation, and that BR is the most susceptible. To illustrate the results from this study, Table 7.1 lists the percentage of voting situations that are susceptible to Coalitional Manipulation for voting rules with 21 voters under the assumption of IC.

Table 7.1 Percentage of voting situations subject to coalitional manipulation with IC

Voting rule	Likelihood of coalitional manipulation (%)
PR	81.0
BR	96.0
PER	23.3
NPER	95.7

The propensity for BR to be manipulated is very high in Table 7.1 at 96.0%, but it is comparable to NPER. The general speculation that BR is particularly sensitive to manipulation is largely supported by these results. Other secondary factors were also examined by Chamberlin, such as the average minimum number of voters that are required for a coalition to successfully use strategic voting with each voting rule. As this minimum required number of voters in a coalition increases, it would become more difficult to coordinate a successful attempt at manipulation. It is found that BR generally holds a middle-ranked position among voting rules, being neither the worst nor the best performer, based on these secondary factors.

The susceptibility of BR, PR and RV to Individual Manipulation was considered in Nitzan (1985) via Monte Carlo simulations based on the assumption of IC. The results that were obtained consistently show that the vulnerability of all three voting rules to individual manipulation increases with the number of candidates and decreases with the number of voters. For any given combination of number of voters in the range $5 \leq n \leq 90$ and number of candidates in the range $3 \leq m \leq 5$, RV is found to be more sensitive to manipulation than BR, which in turn is more susceptible to manipulation than PR. This result is at least valid for this range of a relatively small numbers of voters. Evidence has been found later to indicate that BR is not necessarily more vulnerable to Individual Manipulation than PR when the number of voters is sufficiently large.

A related study was conducted by Smith (1999) that introduces additional measures of Individual Manipulation. These measures account for the number of different voters who could each manipulate the election outcome of voting situations and the amount of improvement that these individual voters could achieve through successful manipulation. The analysis is performed with Monte Carlo simulations based on the assumption of IC, and the results confirm the generally poor performance of BR for most of these measures. The only scenario under which BR performed well in this study was for the very implausible strategy that has voters misrepresent their preferences by randomly selecting some other preference ranking. Just as in the study above by Nitzan (1985), these results apply to elections with a small numbers of candidates and voters.

A series of studies by Aleskerov et al. (2011, 2012, 2015) considers the degree of manipulability of a set of voting rules that includes BR. This analysis behind these studies is slightly different than other studies of strategic voting, since it allows voting rules to be *multiple-choice* in the sense that they are not required to select a single winner by resorting to the use of some tie-breaking mechanism. All of their results come from Monte Carlo simulations that are based on both IC and IAC. It turns out that taking multiple-choice into account does not significantly modify the hierarchy of the voting rules: BR remains among the group of voting rules with a rather high degree of susceptibility to Coalitional Manipulation. However, its performance is somewhat better when Individual Manipulation is considered, since BR is found to perform better than PR in that case.

A perverse voting effect that is very similar in nature to the example in the introduction in which manipulation of BR backfired for the coalition of voters is observed in Laslier (2010) for Individual Manipulation. The study uses Monte Carlo simulation analysis to compare strategic aspects of PR, BR, AV and Copeland Rule (*CR*), which evaluates candidates with a score that measures the number of other candidates that it beats by PMR minus the number of other candidates that beat it. The voters do not know the particular preferences of other voters, but they can respond to a public signal that takes the form of an announced ranking of the candidates by the voting rule that is being used, where ties have been broken randomly. The allowable manipulation strategy that is available for any voter is specifically fixed. To illustrate this allowable strategy for BR, suppose that the announced ranking on four candidates $\{A, B, C, D\}$ is ***ABBBCBD***, but the specific Borda Scores for the candidates are not revealed. So, it is not known if a tie exists between Candidates *A* and *B* with the Borda Scores.

Suppose that a given voter has preferences that agree with this ranking with $A \succ B \succ C \succ D$ and this voter wants to give some additional support to Candidate *A*, just in case it is tied with *B* based on the Borda Scores. Following the logic of the example in the introduction, the voter does this by switching the second-ranked *B* with the last-ranked *D* to manipulate the reported ranking to $A \succ D \succ C \succ B$. By following the same principle, a voter might have the actual preference ranking $B \succ A \succ C \succ D$ and think that Candidates *B* and *A* could be tied based on Borda Score. The voter manipulation in this case would also switch the second-ranked and least-preferred candidates to obtain $B \succ D \succ C \succ A$.

It is then shown that the use of this strategy by either voter could possibly lead to a substantial perverse effect from the attempted Individual Manipulation. This is the same effect that was observed in our example in the introduction when Coalitional Manipulation backfired with BR, where we now find that the least-preferred Candidate *D* would be elected by BR as a result of this strategic voting. The study then goes on to show that PR, AV and CR are all immune to the possibility of displaying this effect.

Analytical representations for the probability that Individual and Coalitional Manipulation can occur for BR and CR with IAC are obtained in Favardin et al. (2002). BR is consistently found to be more vulnerable to manipulation than CR in both cases for any number of voters. To illustrate these results for the case of 51 voters, the vulnerability to Individual Manipulation is 2.1% for CR and 3.8% for BR; and the vulnerability to Coalitional Manipulation is 22.7% for CR and 47.9% for BR. A significant component of our earlier analysis has focused on the general propensity of the introduction of a degree of dependence among voters' preferences with IAC to reduce the likelihood that paradoxical outcomes are observed. We do the same thing here and note that the probability that BR is susceptible to Coalitional Manipulation with IAC for 21 voters is given in Favardin et al.

(2002) as 44.8%. This is dramatically smaller than the associated IC-based probability of Coalitional Manipulation with BR of 96.0% in Table 7.1, but the probability that BR can be manipulated still remains quite high with IAC. It was later shown by Wilson and Pritchard (2007) that the vulnerability of BR to Coalitional Manipulation for the limiting case as $n \to \infty$ under IAC tends to 50.25%.

All of the evidence to this point is making the case against BR look quite strong, since it has consistently been found to have a greater propensity to be manipulated than the other voting rules that it has been compared to. The long-standing conventional belief that BR is particularly susceptible to manipulation has definitely been supported. However, it is very important to recall the events of Chap. 5, when the case against BR looked very strong when it was being compared to PER based on the criterion of Condorcet Efficiency. We have found so far that BR has many good properties, so we continue on with the hope of once again being able to conclude that "The report of the demise of BR is exaggerated."

7.2.2 BR: Not Really All that Bad . . . and Sometimes It Is Optimal

The first glimmer of hope that BR might not be particularly vulnerable to strategic voting comes from Peleg (1979), which demonstrates with a generalized form of IC that the probability that Individual Manipulation is possible tends to zero for every $WSR(\lambda)$ in the limiting case of voters as $n \to \infty$. The study specifically concludes that no $WSR(\lambda)$, which would include BR, should experience any significant problems from Individual Manipulation with large electorates.

Baharad and Neeman (2002) then go on to extend and generalize this result to consider a number of different probability models that describe how voting situations are obtained, including models that allow for some degree of "localized correlation" among the voters' preferences. The probability that a small coalition of voters can successfully manipulate any $WSR(\lambda)$ is shown to decrease to zero as the total number of voters increases, to establish an overall "asymptotic strategyproofness" for all $WSR(\lambda)$ as $n \to \infty$. The same result is found to be valid for voting rules that must elect the CW, when there is one. But, this result does not hold for some of these Condorcet consistent rules when a CW does not exist. The analysis is concluded by stating that if voters must incur some small cost to determine how a voting rule could possibly be manipulated, then all $WSR(\lambda)$ and some Condorcet consistent rules will be highly resistant to strategic manipulation with large electorates. They go on to specifically state that these particular voting rules can be expected to perform very well "whether people are honest or not." This final comment apparently refers back to Borda's statement that was mentioned earlier regarding his voting scheme being intended for use only by "honest men." In another study, Slinko (2002) obtains a similar result to extend this conclusion to two-stage runoff procedures.

A more elementary version of manipulation is considered in Saari (1990), where the objective of manipulation is not necessarily to change the winner of an election, but just to modify the overall ranking of candidates that is obtained with a voting rule. In particular, manipulation is assumed to occur if strategic voting can cause two candidates to switch their positions in the outcome ranking. A micro-manipulation is said to occur in this context if it can be accomplished through a coordinated effort by a coalition that comprises only a small percentage of the electorate. The bounds on the size of the small percentage of voters in such a coalition is not specifically analyzed in the paper, but it is indicated that the proven results will hold when the group of strategic voters represents less than 5% of the electorate. He considers all WSR(λ) and a class of "multiple voting systems", to show that the voting rule that is least susceptible to micro-manipulation for three candidates is BR, for a sufficiently large number of voters under a slightly modified UC assumption. The optimal voting rule changes for a larger number of candidates, but the analysis shows that BR will always perform well anyway. That is, even when BR is not the best choice, it is not very far from being so.

These same results are then extended to consider uncoordinated macro-manipulations in which many small groups of voters attempt to use strategic voting, but where they do so without any coordination of their efforts. It is concluded that such an uncoordinated effort by groups to modify preference rankings on candidates will result in a cancellation of the strategic efforts between opposing groups. After all of these offsetting cancellations are accounted for, the problem reduces to an **equivalent** micro-manipulation problem. Saari also stresses that it is possible to justify any given voting rule in this fashion by appropriately picking some measure of susceptibility to manipulation and then selectively imposing assumptions to restrict how voters' preferences are generated.

Pritchard and Slinko (2006) revisit the notion of evaluating voting rules based on the average minimum size that is required for a coalition to be capable of successfully manipulating each of these rules. This measure was introduced earlier in our discussion of Chamberlin (1985), and the greater the value of this characteristic is for a voting rule, the more capable that rule is to avoid manipulation. It is then shown that this measure of susceptibility to Coalitional Manipulation is not particularly meaningful for NPR, and they consequently suggest the use instead of a more technical measure that is defined as the average minimum size of a "threshold coalition." Both of these measures coincide for all $WSR(\lambda)$ for the limiting case of voters as $n \to \infty$, and it is then proved for that case with IC that BR is the optimal $WSR(\lambda)$ to maximize the average minimum size of a threshold coalition for three and four-candidate elections.

Moyouwou and Tchantcho (2017) develop representations for the limiting IAC probability that BR, PR and NPR are susceptible to strategic manipulation with the additional restriction that the proportion of voters in a manipulating coalition is limited to be p or less. The very important result is found that BR is the least susceptible of the three voting rules to manipulation when p is less than .1058. Secretly coordinating manipulation by a greater proportion of voters with a large electorate would be impossible, and this will soon be seen to be a very important factor.

A	A	B	C	B	C
B	C	A	A	C	B
C	B	C	B	A	A
$n_1 = 3$	$n_2 = 0$	$n_3 = 2$	$n_4 = 0$	$n_5 = 0$	$n_6 = 0$

Fig. 7.4 Sincere preferences for an example of strategic reaction

The tide is definitely starting to turn in favor of BR, since these immediately preceding studies are refuting the commonly held belief that BR is particularly sensitive to Strategic Voting. And, the most conclusive piece of evidence to support BR is presented in Favardin and Lepelley (2006), where the logic behind the assumption of naïve voters is finally brought into question. Why would all other voters idly sit by, and not respond to the possible threat that they could be receiving a less-desirable outcome from an election because some coalition of voters could be utilizing strategic voting against them?

When voters are *non-naïve*, the strategic behavior of any voter can possibly be neutralized by the actions of others, as suggested in the discussion of the cancellation of strategic efforts by competing coalitions above from Saari (1990). They introduce the notion of a *strategic reaction*, which occurs when a voter, or a coalition of voters, prefers the outcome that would result from sincere voting; and where they could possibly be harmed from a threat of strategic voting by another voter, or another coalition of voters. Their response to this threat would be to react strategically with their own voting, in order to increase the probability that the sincere outcome will prevail. We illustrate this process with BR in a three-candidate election by starting with the voting situation in Fig. 7.4 that represents the sincere preferences for five voters.

These sincere preferences result in a situation where Candidate A is the CW, with AMB (3-2) and AMC (5-0). The Borda Scores are calculated as:

$$Score(A) = 1(3) + \frac{1}{2}(2) + 0(0) = 4.0,$$

$$Score(B) = 1(2) + \frac{1}{2}(3) + 0(0) = 3.5,$$

$$Score(C) = 1(0) + \frac{1}{2}(0) + 0(5) = 0.$$

So, Candidate A is selected as the winner with these sincere preferences. The two voters with preferences $B \succ A \succ C$ really want B as the winner, and they choose to behave in the same fashion as the two conniving friends in the example from the introduction, to instead report their preferences as $B \succ C \succ A$, to produce the manipulated voting situation in Fig. 7.5.

The resulting Borda Scores from this strategic voting would be:

Fig. 7.5 Manipulated preferences for an example of strategic reaction

A	A	B	C	B	C
B	C	A	A	C	B
C	B	C	B	A	A
$n_1 = 3$	$n_2 = 0$	$n_3 = 0$	$n_4 = 0$	$n_5 = 2$	$n_6 = 0$

Fig. 7.6 Manipulated preferences with a strategic reaction

A	A	B	C	B	C
B	C	A	A	C	B
C	B	C	B	A	A
$n_1 = 0$	$n_2 = 3$	$n_3 = 0$	$n_4 = 0$	$n_5 = 2$	$n_6 = 0$

$$Score(A) = 1(3) + \frac{1}{2}(0) + 0(2) = 3.0,$$

$$Score(B) = 1(2) + \frac{1}{2}(3) + 0(0) = 3.5,$$

$$Score(C) = 1(0) + \frac{1}{2}(2) + 0(3) = 1.0.$$

Candidate B would therefore be the winner as a result of the manipulation by the two voters. But, the three voters with preferences $A \succ B \succ C$ are not naïve in this case, and they are aware of the possibility that the two voters could manipulate their preferences to obtain the voting situation in Fig. 7.5. They really want their most-preferred Candidate A to be the winner, as it would have been with sincere voting. So, they strengthen the Borda Count for A over B with the strategic reaction by misrepresenting their own preferences as $A \succ C \succ B$, to arrive at the voting situation in Fig. 7.6.

The Borda Scores are now calculated to be:

$$Score(A) = 1(3) + \frac{1}{2}(0) + 0(2) = 3.0,$$

$$Score(B) = 1(2) + \frac{1}{2}(0) + 0(3) = 2.0,$$

$$Score(C) = 1(0) + \frac{1}{2}(5) + 0(0) = 2.5.$$

Candidate A has now re-emerged as the BR winner, just as it was with sincere voting. Note that the Borda Score for Candidate C has increased from 1.0 with sincere voting to 2.5 due to the combined effects of the initial manipulation and the strategic reaction to it. The process worked in this example, to reinstate the winning candidate with BR for sincere preferences; but it could have backfired if the Borda Score for C had increased any more than it did, to elect every voter's least preferred candidate with sincere preferences. The final outcome of this action and reaction resulted in no change from the winning candidate with sincere preferences, so the attempt at manipulation was meaningless because the voters were non-naïve. It turns out that effectively no manipulation occurred as a result, to make this a "quasi-stable" situation.

Favardin and Lepelley (2006) make a further distinction regarding how manipulation might occur. The first case represents *Homogeneous Voters*, which is similar to all of the examples that we have considered so far. That is, all of the voters included in a manipulating coalition have identical preferences rankings on the candidates. This case would be most relevant to scenarios like a political assembly, where the set of all members can be neatly partitioned into several groups according to party membership. Homogeneous preferences would be expected to prevail within each of these groups, with preferences being aligned according to the associated party ideologies. Such a scenario would typically be very amenable to communication within each of the individual groups of homogeneous voters before an election, to lead to the development of insular voting strategies for each group.

Extending the same level of communication across different groups to lead to a coalition of groups that agree to cooperatively adopt strategies that are mutually beneficial for each of these groups might be more challenging to achieve in an assembly. Such a coalition of groups would be composed of *Heterogeneous Voters*, and the possibility of across-group cooperation obviously allows for more options for Coalitional Manipulation to occur. Representations are obtained to yield limiting IAC probabilities as $n \rightarrow \infty$ that a number of voting rules are susceptible to Individual and Coalitional Manipulation with both naïve and non-naïve voters. We focus on the differences in the results that were obtained for the more interesting case of Heterogeneous Voters, since that scenario results in a greater potential level of susceptibility to manipulation.

When Individual Manipulation was considered for naïve voters, BR was found to be one of the set of voting rules with the lowest susceptibility to manipulation for all cases with more than 15 voters, which is in complete agreement with the expectations from Saari (1990) that were discussed above. The computed probabilities of susceptibility to Coalitional Manipulation with naïve voters are shown in Table 7.2 for PR, NPR, BR, PER and NPER. These results for the voting rules that we have given the most attention to were accumulated from Lepelley and Mbih (1987, 1994) and Wilson and Pritchard (2007), where it is seen that BR is second only to NPR for having the highest likelihood of manipulation. This limiting probability estimate of .5025 for BR was mentioned earlier as strongly reflecting an overall poor performance.

An extraordinary outcome happens in Favardin and Lepelley (2006) when they instead consider the results for non-naïve voters. As is the case with naïve voters,

Table 7.2 Limiting probability of coalitional manipulation for heterogeneous voters with IAC

Voting rule	Naïve voters	Non-naïve voters
PR	.2916	.1736
NPR	.5185	.4444
BR	.5025	.1375
PER	.1111	.0920
NPER	.4306	.0903

BR remains among the set of voting rules that have the lowest potential for Individual Manipulation. The probabilities of susceptibility to Coalitional Manipulation with non-naïve voters are listed in Table 7.2 for PR, NPR, BR, PER and NPER, where BR shows an absolutely remarkable decrease in its manipulation probability to .1375, and BR is now clearly superior to both PR and NPR. Both PER and NPER do perform better than BR, but these results strongly indicate that, contrary to common belief, BR is not at all particularly vulnerable to manipulation when non-naïve voters are considered. It is therefore very natural to wonder if voters really do tend to react to potential threats of manipulation in a non-naïve fashion.

Some evidence to provide an answer to this question comes from an empirical laboratory study by Béhue et al. (2009), where evidence is found to support the premise that voters do indeed react to potential threats of manipulation by acting like non-naïve voters. Two types of voting situations with three candidates are compared in this study. In the first scenario, Type 1 voting situations are contrived to have one voter who can successfully manipulate the election, and where no possible reaction to this threat exists for any of the other voters. In Type 2 voting situations there is still one voter who is capable of manipulating the election, but now there is also another voter who can use a strategic reaction to that threat to restore the original winner from sincere voting. The results show that the sincere winner is elected with BR in only 33% of the elections that were conducted with Type 1 situations, so the voters who can manipulate the outcome without any interference clearly are taking advantage of that opportunity. And, the sincere winner is elected in 75% of the elections with Type 2 situations, so that the voters who have the possibility of using a strategic reaction are frequently taking advantage of that option when it is available. This provides some credible support for the premise that voters generally do tend to act in accordance with the model of non-naïve voters when a threat of possible manipulation is present.

7.2.3 Other Studies of BR Manipulation

A laboratory experiment was used in Kube and Puppe (2009) to study the overall propensity of voters to engage in strategic voting with BR for three voters and four candidates. Different behavior patterns were found to exist for voters, based on whether, or not, an individual voter knew that other voters were using strategic manipulation. In each election, one single player was provided with additional information about the other players' preference rankings on the candidates. It is found that rates of attempts to manipulate outcomes are surprisingly low, even for the individual voters who possessed superior information about the other voters' preferences. However, when this individual voter was provided with both the other voters' preferences and information about how they actually voted, their manipulation rates increased significantly. This rate rose to about 50% in one scenario, to suggest that uncertainty about how other voters might be behaving plays a key role in understanding the use of strategic voting in elections. It is asserted that this particular manipulation rate for such a highly contrived scenario can only be viewed as an upper bound for actual elections. That is, typical elections that are conducted under natural conditions that are both more complex and also involve voters who are inexperienced with the complexities of manipulation should be expected to have much lower rates of strategic voting. Similar conclusions can be found in a number of other studies that attempt to measure the percentage of strategic voters in real elections, where it is found that typically only around 5% of voters do so [see for example Blais (2002)].

The final study that we include regarding the susceptibility of BR to strategic voting takes a very different and interesting approach by starting with the observation that the manipulation of a voting rule is not necessarily a "bad" thing. As an elementary example, Coalitional Manipulation can prevent the election of a CL with PR! This line of reasoning is pursued in Lehtinen (2007) to evaluate strategic voting under BR in the context of the overall expected value of utility that is obtained for the electorate. The analysis is performed with Monte Carlo simulation analysis for three-candidate elections.

The model that is used is based on the utilities that each voter has for the candidates, and voters' preference rankings are obtained accordingly. Some incomplete information about the sincere preferences in the voting situation of the electorate is available from "noisy signals" before the election is held, and voters can use this information to derive their beliefs about whether, or not, one candidate has a higher Borda Score than another. Expected Utility Behavior is then defined to permit voters to vote either strategically or sincerely, depending on their preferences and beliefs; and Utilitarian Efficiency is defined as the percentage of simulated elections in which the candidate with the highest sum of utilities over all voters is elected. The main finding of the study is that the use of strategic voting with Expected Utility Behavior yields higher Utilitarian Efficiencies than sincere voting does. As a result of this, the arguments that have previously been thought of as being a major drawback from using BR actually turn out to be an argument to support using it, when attention is shifted to focus on maximizing overall utility. Lehtinen remarks that strategic voting actually increases Utilitarian Efficiency for most commonly used voting rules. But, BR has two advantages over some of the other rules. Strategic voting increases overall utility, even if the voters have unreliable information regarding other voters' preferences. And, BR tends to yield high Utilitarian Efficiencies, even if only a few voters engage in strategic behavior.

Based on all of this information, a variety of arguments exist to defend BR against the commonly held perception that it is particularly susceptible to strategic voting. First, the frequency of any actual success with strategic voting is found to be low in real elections for various voting rules that are used in practice, including BR. Second, when manipulation can be accomplished, it turns out that BR is less susceptible to manipulation than other voting rules in a number of different scenarios, particularly with non-naïve voters. The conclusion that BR is less susceptible than other $WSR(\lambda)$ is definitely quite robust. Third, the impact from successfully using strategic voting in an election actually is not necessarily a negative outcome with BR, since its manipulation can be expected to result in an increase in the likelihood of electing the overall utility-maximizing candidate. We find that it is indeed true that "The report of the demise of BR is exaggerated."

7.3 The Three-Scale Evaluative Voting Rule

Three-Scale Evaluative Voting (*EV*) is similar in nature to AV, where AV is implemented by having voters assign a weight of one to each candidate that they approve of, and a weight of zero to candidates that they do not approve of. It is clear that AV was designed with the basic premise that voters can adequately represent their preferences on candidates by partitioning them into two sets that are represented by dichotomous preferences, and this was thoroughly examined in Chap. 6. The basic notion behind EV allows for a refinement of the AV assumptions, such that voters' preferences can take on three levels of acceptability to be trichotomous. Voters will then assign a weight of two to candidates that are considered to be in a preferred group, one point to candidates that voters view with indifference and zero points to candidates that are disapproved of. The placement of candidates into the indifference or neutral category could result from voters having insufficient information to form a distinct opinion about them. Voting with EV has received a lot of recent attention, and it is considered to have some nice properties [see for example Smaoui and Lepelley (2013)]. Alcantud and Laruelle (2014) develop the necessary and sufficient properties that characterize EV.

It has been mentioned before that it is difficult to explore any aspects of voting theory without finding footprints that were left behind by Condorcet, and EV is not an exception. A voting rule was suggested by Condorcet (1793) for use in electing groups of representatives from districts to the National Assembly, as part of a proposed new constitution for France. There were to be three times as many nominated candidates as the number of seats to be filled in each district. Voters were then required to partition these candidates into three groups of equal size according to their preferences, and present a ballot that listed the most preferred group of Election Votes and the middle-ranked group of Supplemental Votes. This particular requirement that the number of candidates that are placed into each group must be equal varies from our definition of EV. Candidates who were listed on a majority of ballots among Election Votes were then chosen in order, according to the number of votes received. If not enough candidates received a majority of votes on the ballots in this first step, the Supplemental Votes were then included with the Election Votes in a second step, to look for candidates who appeared on a majority of ballots in the combined set. As mentioned in the first chapter, this proposed constitution was defeated and Condorcet died in prison soon afterward. However, this particular voting rule was actually adopted for use in Geneva. So far, this does not describe EV, except for the partitioning of candidates into three categories. A voting rule more closely related to EV came when this rule was modified later.

After Condorcet's procedure had been used in Geneva, Lhuilier (1794) wrote about some difficulties that had been observed with its use, and suggested a slightly modified procedure. An interesting sidelight is that he mentions the fact that Condorcet could not be contacted regarding these newfound results. During this period of very limited communication possibilities in France, Lhuilier apparently was completely unaware that Condorcet had been imprisoned and died there. The modification that was suggested applied to the second step of the process, such that the number of Supplemental Votes for candidates would only receive one-half of the weight that was given to the Election Votes. Lhuilier shows that using these BR weights for the three categories leads to a significant improvement in the results that are obtained in three-candidate elections, compared to using Condorcet's proposed method with equal weights for both groups, and this modified procedure is definitely more like EV. It was further suggested that different weights might be more appropriate for use in situations with more than three candidates.

It was mentioned above that EV has received a lot of recent attention that has been largely focused on showing that it has some very nice properties, but some potential difficulties with using this voting rule have also been pointed out in Smaoui and Lepelley (2013) and Felsenthal (2012). In particular, it has been noted that EV does not necessarily elect the CW, and it can exhibit a Strong Borda Paradox by electing the CL. This concern was investigated in El Ouafdi et al. (2017a) that extended the earlier analysis that compared AV to other voting rules under the condition that voters have dichotomous preferences. They instead compared EV to other voting rules for the case of three candidates when voters have trichotomous preferences.

The definition of EV was specifically based on the assumption of trichotomous preferences, but the other voting rules must be modified to account for this scenario. This is accomplished in a straightforward manner, based on how the candidates are distributed among the preferred, indifference and disapproved categories. Voting situations for which all three candidates are contained in the same preference category are ignored, since this effectively indicates that a voter is indifferent between all three candidates. If the candidates are partitioned into just two categories, the empty category is ignored and the remaining two categories are treated like dichotomous preferences. If there is a candidate in each of the three categories, a linear preference ranking exists. Then, the Extended Weighted Scoring Rule definitions for EPR, ENPR and EBR can be used for the cases in which candidates are partitioned into two or three preference categories. If the candidates are partitioned into two categories, then AV can be directly applied. Some accommodation must be made for AV if a linear preference ranking exists, and this is done in El Ouafdi et al. (2017a) by assuming in this case that it is equally likely that the middle-ranked candidate in a voter's preference ranking will be approved or disapproved with AV.

Limiting representations were then obtained for large electorates as $n \to \infty$ with IAC for both the Condorcet Efficiency and the Strong Borda Paradox Probability that the CL is elected for each of EV, AV, EPR, ENPR and EBR. The results are listed in Table 7.3.

Table 7.3 IAC efficiencies and Borda Paradox probabilities with trichotomous preferences

Voting rule	Condorcet Efficiency	Borda Paradox probability
EV	.8652	.0098
EPR	.8487	.0223
ENPR	.7821	.0217
EBR	.9339	.0000
AV	.9153	.0011

These results indicate that EV outperforms both EPR and ENPR on the basis of both Condorcet Efficiency and the probability that a Strong Borda Paradox is observed. It is also seen that AV outperforms EV on both factors, but different results were obtained in Smaoui and Lepelley (2013) where a different set of assumptions that will be discussed later were used. The most important finding is that the consistent pattern emerges again, where BR is found to be superior to the other single-stage voting rules both on the basis of Condorcet Efficiency and the probability that a Strong Borda Paradox will be observed.

7.4 The Case of More than Three Candidates

We have focused almost completely on the outcomes from three-candidate elections, to conclude that BR looks like a very good option for use in an election, but it is obvious to wonder how robust this conclusion might be when more candidates are considered in an election. The process of obtaining exact limiting probability values with IC and IAC as $n \to \infty$ becomes quite restricted when the case of more than three candidates is being evaluated. However, significant progress has been made in this area with IAC for the case of three and four candidates.

For the case of four candidates in the limiting case for voters as $n \to \infty$, Schürmann (2013) obtains the Condorcet Efficiency of PR. Other results from El Ouafdi et al. (2017b) do the same for of BR and NPR, and representations for the Condorcet Efficiency of two-stage elimination procedures are also obtained. For the case of more than three candidates, these runoff procedures retain only the two top-scoring candidates for the second stage of voting by majority rule. The runoff procedures are based on using the first-stage voting rules of Plurality Rule (PRR), Negative Plurality (NPRR) and Borda Rule (BRR). The results in Table 7.4 summarize the computed values from these representations for four candidates, and the results for the three-candidate cases from Tables 4.1 and 5.1 are included for comparison. The runoff voting rules PRR, NPRR and BRR are replaced by PER, NPER and BER for the case of three candidates.

It is interesting to note that the reduction in the rate of the Condorcet Efficiency that occurs when moving from three to four candidates with IAC is lower for BR than it is for PR, NPR, PRR and NPRR. The efficiency for BR is now closer to the two-stage PRR value, and it is higher than the NPRR value! An independent study

Table 7.4 Limiting Condorcet Efficiency with IAC

Voting rule	Three candidates	Four candidates
PR	.8815	.7426
NPR	.6296	.5516
BR	.9111	.8706
PRR	.9685	.9116
NPRR	.9704	.8450
BRR	1.0000	.9961

Table 7.5 Limiting strong Borda Paradox probability with IAC

Voting rule	Three candidates	Four candidates
PR	.0296	.0227
NPR	.0315	.0238
BR	.0000	.0000

by Bruns et al. (2017) confirms the computed value for Condorcet Efficiency of PRR with four candidates. These results strongly reinforce our conclusion in Chap. 5 that the expected benefit that would be gained from using two-stage voting rules like PRR or NPRR instead of BR is quite small.

The limiting probability of observing a Strong Borda Paradox with PR, NPR and BR for four candidates with IAC is also considered in El Ouafdi et al. (2017b). The results are listed in Table 7.5, along with their corresponding values for three candidates from Table 2.9.

The study by Bruns et al. (2017) confirms the probability that the CL is elected by PR with four candidates, and the probability of observing a Strong Borda Paradox decreases for both PR and NPR in the move from three to four candidates.

When more than four candidates are considered, Monte Carlo simulation estimates are typically used in studies of the Condorcet Efficiency of voting rules. Lepelley et al. (2000) performed such an analysis to obtain estimates of the limiting efficiencies of PR, NPR and BR with IC and IAC for the number of candidates in the range $3 \leq m \leq 8$, and the results are summarized in Table 7.6.

Based on earlier discussion, it is not at all surprising to see that the IC and IAC estimates for these Condorcet Efficiencies converge to very similar values as m increases for each voting rule. What is surprising is the degree to which BR significantly dominates PR and NPR as m increases. As the number of candidates increases from three to eight, PR efficiency drops from .8816 to .4088 with IAC, which is a 53.6% decrease from the starting value with three candidates. On the other hand, BR drops from .9108 to .8428, which corresponds to only a 7.5% decrease from its starting value with IAC! The Condorcet Efficiency dominance of BR in Table 7.6 is blatantly obvious as the number of candidates increases for these single-stage voting rules.

Another Monte Carlo simulation-based analysis of the Condorcet Efficiency of voting rules with more than three candidates was conducted by Smaoui and Lepelley (2013), and it is designed to allow for a more natural comparison of AV and EV to other voting rules. Voting situations are developed on a utility-based

Table 7.6 Limiting Condorcet Efficiency estimates with IC and IAC

	Voting rule					
	PR		NPR		BR	
m	IC	IAC	IC	IAC	IC	IAC
3	.7574	.8816	.7571	.6298	.9010	.9108
4	.6416	.7429	.6415	.5517	.8702	.8706
5	.5570	.6139	.5602	.5090	.8552	.8541
6	.4858	.5198	.4946	.4730	.8450	.8471
7	.4663	.4524	.4450	.4386	.8438	.8457
8	.4123	.4088	.4378	.4101	.8362	.8428

Table 7.7 Condorcet Efficiencies for six voting rules with a utility based model

m	EV	PR	PRR	NPR	BR	AV
3	.7066	.7600	.8987	.7278	.9005	.6006
4	.6699	.6376	.8002	.5960	.8719	.5390
5	.6481	.5491	.7121	.5020	.8532	.4981
6	.6322	.4831	.6430	.4317	.8391	.4678
7	.6209	.4308	.5846	.3775	.8278	.4451
8	.6118	.3893	.5360	.3348	.8190	.4257

model for obtaining preferences for each voter. For each individual voter, a random utility value is assigned to each of the m candidates from a uniform distribution over the interval [0, 1]. Voting with PR, PRR, NPR and BR is then conducted by using the preference rankings for voters that follow from the ordinal rankings of candidates according to decreasing values of their utilities, which is consistent with IC. The CW was also determined from these ordinal rankings in all cases. When voting with AV, each voter approves of all candidates for which their utilities are in the range [1/2, 1], and disapproves of candidates with utilities in the range [0,1/2]. When voting with EV, each voter categorizes all candidates with a utility in the range [1,2/3] as being preferred, all candidates with utilities in the range [2/3,1/3] as being in the indifference group and all candidates with utilities in the range [1/3,0] as being disapproved.

The study was based on $n = 300$ voters for samples of 10,000 voting situations for which a CW was found to exist for each case, with the number of candidates in the range $3 \leq m \leq 8$. The results are summarized in Table 7.7.

The results of this utility-based analysis once again show that BR consistently dominates all other voting rules on the basis of Condorcet Efficiency, including the two-stage PRR. Moreover, the Condorcet Efficiency of BR remains rather high when the number of candidates increases, in contrast with PRR. We find that EV now dominates AV with this model, as opposed to what was observed with three-candidates in the results from the scenario that required voters to have trichotomous preferences in Table 7.3.

Table 7.8 Limiting IC probabilities and efficiencies

α_{PP}	Actual = Observed	PR & NPR efficiency	BR efficiency
≈0	.3041	1/3 = .3333	1/3 = .3333
.1	.4236	.4399	.4576
.2	.4806	.4882	.5151
.3	.5290	.5277	.5630
.4	.5742	.5630	.6068
.5	.6186	.5961	.6490
.6	.6640	.6280	.6911
.7	.7126	.6595	.7345
.8	.7675	.6911	.7810
.9	.8365	.7234	.8337
1.0	1.0000	.7572	.9012

7.5 Final Conclusion: The Impact of Abstentions

The consistent message that has been received in this analysis is that BR really does perform very well, relative to other voting rules, on the basis of Condorcet Efficiency; and that the commonly held belief that BR is particularly sensitive to strategic voting is not really supported. So, what remains to be considered? The answer to this question is that the impact of abstention rates can be a critical factor in evaluating the effectiveness of voting rules. Recall what was observed during the discussion of the Forced Ranking Option, when it became very clear that there were very large differences between what happened when the Actual and the Forced Winners were considered. These differences were so glaring that it became obvious that the Forced Ranking Option should never be used.

We still define the Actual Winner based on the true preferences of the entire electorate of possible voters. When abstention is an option, we instead define the Observed Winner solely on the basis of the preferences of the subset of voters from the entire electorate who choose to participate. Gehrlein and Fishburn (1978) use an IC-based assumption $IC(\alpha_{PP})$ for three candidates to develop a representation for the limiting conditional probability that the Actual and Observed Winners coincide, given that an Actual Winner exists. Each of the six linear preference rankings is assumed to be equally likely to represent the preferences of any voter with this model, and the probability that each potential voter participates in the election is α_{PP}. Computed results from this representation are shown in Table 7.8.

Some results from Table 7.8 are quite predictable. The Actual and Observed Winners must obviously coincide if $\alpha_{PP} = 1$. Nothing is known if no voters participate in the election, but the .3041 probability that the Actual and Observed Winners coincide as $\alpha_{PP} \rightarrow 0$ is 1/3 of the limiting probability that a CW exists with IC that we considered earlier.

Gehrlein and Fishburn (1979) use the same model to determine limiting representations for the Actual Efficiency of a general $WSR(\lambda)$ and find that the PR and NPR efficiencies are identical. Computed values are obtained from this

Table 7.9 Limiting IAC probabilities and efficiencies

α_{PP}	Actual = Observed	PR efficiency	NPR efficiency	BR efficiency
≈0	5/16 = .3125	1/3 = .3333	1/3 = .3333	1/3 = .3333
.1	.3482	.3679	.3621	.3704
.2	.3949	.4131	.3975	.4184
.3	.4564	.4723	.4411	.4807
.4	.5357	.5481	.4928	.5600
.5	.6310	.6378	.5475	.6533
.6	.7291	.7267	.5912	.7457
.7	.8160	.7988	.6153	.8212
.8	.8887	.8476	.6255	.8737
.9	.9492	.8738	.6289	.9026
1.0	1.0000	119/135 = .8815	41/45 = .6296	17/27 = .9111

representation for PR, NPR and BR in Table 7.8. As $\alpha_{PP} \to 0$, the voting rules are found to approach random selection procedures, with Actual Efficiencies equal to 1/3. As $\alpha_{PP} \to 1$ the efficiencies approach the known IC values when no abstention is allowed, from Table 4.1. All of the results in Table 7.8 show what has by now come to be expected. That is, BR outperforms the other voting for all levels of voter participation, but a very unfortunate part of this conclusion is that the BR Efficiencies also become very weak for low participation probabilities.

The only remaining hope is that we can rely on the introduction of some degree of dependence among voters' preferences with IAC to increase these efficiencies to more acceptable levels. That particular aspect of this problem was studied in Gehrlein and Lepelley (2017), where the $IAC(\alpha_{PP})$ assumption was used for the limiting case of voters, where all voting situations that have the same participation proportion α_{PP} of voters from the entire electorate are equally likely to be observed. The results from that analysis are shown in Table 7.9.

As $\alpha_{PP} \to 0$ in Table 7.9, the voting rules approach random selection procedures, with PR, NPR and BR having Actual Efficiency values of 1/3. The probability that the Actual and Observed Winners coincide is 5/16, which is 1/3 of the limiting probability that a CW exists with IAC. As $\alpha_{PP} \to 1$ the efficiencies approach the known IAC values from Table 4.1 where no abstention is allowed.

The results of this study show that the introduction of a degree of dependence among voters' preferences provides some good news, and a lot of bad news, when we compare the results of Tables 7.8 and 7.9. The results with the $IAC(\alpha_{PP})$ model show a uniform decrease in Actual Efficiency for NPR, compared to the $IC(\alpha_{PP})$ results. However, NPR has already been shown to be a consistently poor performer. All of the values of the probability that the Actual and Observed Winners coincide, and the Actual Efficiencies for PR and BR, increase with $IAC(\alpha_{PP})$ for participation rates with $\alpha_{PP} \geq 1/2$, and decrease for $\alpha_{PP} < 1/2$. As in the case with $IC(\alpha_{PP})$, BR consistently outperforms PR and NPR on the basis of Actual Efficiency. However, its "relatively best" performance with $IAC(\alpha_{PP})$ is even worse than we found with

$IC(\alpha_{PP})$ for low levels of voter participation. So, the Actual Efficiency values for BR, PR and NPR are surprisingly found to be lower with $IAC(\alpha_{PP})$ than with $IC(\alpha_{PP})$ for all $0 < \alpha_{PP} < 1/2$! This leaves little doubt that BR tends to be the best voting rule for use to maximize Actual Efficiency, but this matters very little when participation rates are low. While BR might be the best in that case, none of the voting rules are very good at selecting the Actual Winner.

Given this insight, what do actual participation rates look like? DeSilver (2016) lists the percentage of voting-age citizens who participated in recent elections in 35 countries that are members of the Organization for Economic Cooperation and Development. Most member countries of this group are highly-developed democratic states. The participation percentages range from 38.6% in Switzerland for 2015 to 84.3% in Turkey for 2015. The participation rates for Australia in 2016 is listed at 91.0%, and at 87.2% in 2014 for Belgium; but all voting-age citizens in both of those countries are legally obligated to report to a polling place on election days, so they were not included in the range that is reported here.

The implementation of compulsory voting for every eligible citizen sounds like a direct solution to this problem. This approach has been used in a number of different countries in different formats. Citizens can be required to actually vote, or they can only be required to report to a polling place on election days without having to actually vote. There is a wide variation in the penalties that are imposed for non-compliance, and there are major differences in how strictly these laws are actually enforced. All of this is described in detail in Milner et al. (2007) with an empirical analysis that was conducted to determine the impact that implementing compulsory voting would have in Canada, where reported participation rates for disengaged younger eligible voters have consistently declined for many years. The conclusion of that study is that compulsory voting would lead to a significantly increased participation rate, but no evidence was found to lead to the conclusion that the additional voters would be induced to gain any additional knowledge about the issue positions of the political parties before voting. Compulsory voting will get eligible voters to polling places. But, it is also essential that eligible voters are made aware of the importance of voting and of the issue positions that are being taken by the different political parties.

References

Alcantud, J. C., & Laruelle, A. (2014). Dis&approval voting: A characterization. *Social Choice and Welfare, 43*, 1–10.

Aleskerov, F., Ivanov, A., Karabekyan, D., & Yakuba, V. (2015). Manipulability of aggregation procedures in impartial anonymous culture. *Procedia Computer Science, 55*, 1250–1257.

Aleskerov, F., Karabekyan, D., Sanver, R., & Yakuba, V. (2011). On individual manipulability of positional voting rules. *SERIEs: Journal of the Spanish Economic Association, 2*, 431–446.

Aleskerov, F., Karabekyan, D., Sanver, R., & Yakuba, V. (2012). On the manipulability of voting rules: Case of 4 and 5 alternatives. *Mathematical Social Sciences, 64*, 67–73.

Baharad, E., & Neeman, Z. (2002). The asymptotic strategyproofness of scoring and Condorcet consistent rules. *Review of Economic Design, 7*, 331–340.

Béhue, V., Favardin, P., & Lepelley, D. (2009). La manipulation stratégique des règles de vote: Une étude expérimentale. *Louvain Economic Review, 75*, 503–516.

Black, D. (1958). *The theory of committees and elections*. Cambridge: Cambridge University Press.

Blais, A. (2002). Why is there so little strategic voting in Canadian Plurality Rule elections? *Political Studies, 50*, 445–454.

Bruns, W., Ichim, B., & Söger, C. (2017). *Computations of volumes and Ehrhart series in four candidate elections*. Working Paper at www.researchgate.net/publication/315775243_Computations_of_volumes_and_Ehrhart_series_in_four_candidates_elections

Chamberlin, J. R. (1985). An investigation into the relative manipulability of four voting systems. *Behavioral Science, 30*, 195–203.

de Condorcet, M. (1793). A survey of the principles underlying the Draft Constitution presented to the National Convention on 15th and 16th February 1793. In F. Sommerlad, I. McLean (Eds., 1989), *The political theory of Condorcet* (pp. 216–222). Oxford: University of Oxford Working Paper.

DeSilver, D. (2016). *US voter turnout trails most developed countries*. Pew Research Center Report: http://www.pewresearch.org/fact-tank/2016/08/02/u-s-voter-turnout-trails-most-developed-countries/

El Ouafdi, A., Lepelley, D., & Smaoui, H. (2017a). *On the Condorcet Efficiency of evaluative voting (and other voting rules) with trichotomous preferences*. University of La Réunion, Working Paper.

El Ouafdi, A., Lepelley, D., & Smaoui, H. (2017b). *Probabilities of electoral outcomes in four-candidate elections*. University of La Réunion, Working Paper.

Favardin, P., & Lepelley, D. (2006). Some further results on the manipulability of social choice rules. *Social Choice and Welfare, 26*, 485–509.

Favardin, P., Lepelley, D., & Serais, J. (2002). Borda Rule, Copeland method and strategic manipulation. *Review of Economic Design, 7*, 213–228.

Felsenthal, D. (2012). Review of paradoxes afflicting procedures for electing a single candidate. In S. Dan Felsenthal, & M. Machover(Ed.), Electoral systems: Paradoxes, assumptions and procedures (pp. 19–92). Berlin: Springer.

Gehrlein, W., & Lepelley, D. (2017). *The impact of abstentions on election outcomes when voters have dependent preferences*. Technical Note available on Research Gate at: https://www.researchgate.net/publication/316276995_The_Impact_of_Abstentions_on_Election_Outcomes_when_Voters_have_Dependent_Preferences

Gehrlein, W. V., & Fishburn, P. C. (1978). Probabilities of election outcomes for large electorates. *Journal of Economic Theory, 19*, 38–49.

Gehrlein, W. V., & Fishburn, P. C. (1979). Effects of abstentions on voting procedures in three-candidate elections. *Behavioral Science, 24*, 346–354.

Gibbard, A. (1973). Manipulation of voting schemes: A general result. *Econometrica, 41*, 587–601.

Kube, S., & Puppe, C. (2009). When and how do voters try to manipulate? Experimental evidence from Borda elections. *Public Choice, 139*, 39–52.

Laslier, J. F. (2010). In silico voting experiments. In J. F. Laslier & R. Sanver (Eds.), *Handbook of approval voting*. Berlin: Springer.

Lehtinen, A. (2007). The Borda rule is also intended for dishonest men. *Public Choice, 133*, 73–90.

Lepelley, D., Louichi, A., & Valognes, F. (2000). Computer simulations of voting systems. In G. Ballot & G. Weisbuch (Eds.), *Applications of simulations to social sciences*. Oxford: Hermes.

Lepelley, D., & Mbih, B. (1987). The proportion of coalitionally unstable situations under the plurality rule. *Economics Letters, 24*, 311–315.

Lepelley, D., & Mbih, B. (1994). The vulnerability of four social choice functions to coalitional manipulation of preferences. *Social Choice and Welfare, 11*, 253–265.

Lhuilier, S. (1794). An examination of the election method proposed to the National Convention of France in February 1793, and Adopted in Geneva. In F. Sommerlad, & I. McLean (Eds., 1989), *The political theory of Condorcet* (pp. 223–253). Oxford: University of Oxford Working Paper.

Milner, H., Loewen, P. J., & Hicks, B. M. (2007). The paradox of compulsory voting: Participation does not equal political knowledge. *IRPP Policy Matters, 8*. Available at: http://irpp.org/research-studies/policy-matters-vol8-no3/

Moyouwou, I., & Tchantcho, H. (2017). Asymptotic vulnerability of positional voting rules to coalitional manipulation. *Mathematical Social Sciences, 89*, 70–82.

Nitzan, S. (1985). The vulnerability of point-voting schemes to preference variation and strategic manipulation. *Public Choice, 47*, 349–370.

Peleg, B. (1979). A note on manipulability of large voting schemes. *Theory and Decision, 11*, 401–412.

Pritchard, G., & Slinko, A. (2006). On the average minimum size of a manipulating coalition. *Social Choice and Welfare, 27*, 263–277.

Pritchard, G., & Wilson, M. C. (2007). Exact results on manipulability of positional voting rules. *Social Choice and Welfare, 29*, 487–513.

Saari, D. G. (1990). Susceptibility to manipulation. *Public Choice, 64*, 21–41.

Satterthwaite, M. (1975). Strategy-proofness and arrow's condition. *Journal of Economic Theory, 10*, 198–217.

Schürmann, A. (2013). Exploiting polyhedral symmetries in social choice. *Social Choice and Welfare, 40*, 1097–1110.

Slinko, A. (2002). On asymptotic strategy-proofness of classical social choice rules. *Theory and Decision, 52*, 389–398.

Smaoui, H., & Lepelley, D. (2013). Le système de vote par note à trois niveaux: étude d'un nouveau mode de scrutin. *Revue d'Economie Politique, 123*, 827–850.

Smith, D. A. (1999). Manipulability measures of common social choice functions. *Social Choice and Welfare, 16*, 639–661.

Wilson, M. C., & Pritchard, G. (2007). Probability calculations under the IAC hypothesis. *Mathematical Social Sciences, 54*, 244–256.

MIX
Papier aus verantwortungsvollen Quellen
Paper from responsible sources
FSC® C105338
FSC
www.fsc.org

If you have any concerns about our products,
you can contact us on
ProductSafety@springernature.com

In case Publisher is established outside the EU,
the EU authorized representative is:
Springer Nature Customer Service Center GmbH
Europaplatz 3, 69115 Heidelberg, Germany

Printed by Libri Plureos GmbH
in Hamburg, Germany